Basic Concept of Biochemistry of Blood

Basic Concept of Biochemistry of Blood

Kurra Venkata Gopaiah

RANDOM

RANDOM PUBLICATIONS

NEW DELHI (INDIA)

Basic Concept of Biochemistry of Blood

ISBN 978-93-5111-800-8

Published in 2016 in India by

RANDOM PUBLICATIONS

4376-A/4B, Gali Murari Lal, Ansari Road
New Delhi-110 002
Phone : +9111-43580356, 011-23289044, 011-43142548
e-mail: sales@randompublications.com,
info@randompublications.com, randomexports@gmail.com

Reprinted 2026

Type Setting by : Friends Media, Delhi-110089

Preface

Blood fluid that transports oxygen and nutrients to the cells and carries away carbon dioxide and other waste products. Technically, blood is a transport liquid pumped by the heart (or an equivalent structure) to all parts of the body, after which it is returned to the heart to repeat the process. Blood is both a tissue and a fluid. It is a tissue because it is a collection of similar specialized cells that serve particular functions. These cells are suspended in a liquid matrix (plasma), which makes the blood a fluid. If blood flow ceases, death will occur within minutes because of the effects of an unfavourable environment on highly susceptible cells. The constancy of the composition of the blood is made possible by the circulation, which conveys blood through the organs that regulate the concentrations of its components. In the lungs, blood acquires oxygen and releases carbon dioxide transported from the tissues. The kidneys remove excess water and dissolved waste products. Nutrient substances derived from food reach the bloodstream after absorption by the gastrointestinal tract. Glands of the endocrine system release their secretions into the blood, which transports these hormones to the tissues in which they exert their effects. Many substances are recycled through the blood; for example, iron released during the destruction of old red cells is conveyed by the plasma to sites of new red cell production where it is reused. Each of the numerous components of the blood is kept within appropriate concentration limits by an effective regulatory mechanism. In many instances, feedback control systems are operative; thus, a declining level of blood sugar (glucose) leads to accelerated release of glucose into the blood so that a potentially hazardous depletion of glucose does not occur.

*– **Author***

Contents

1

Introduction

Blood is a bodily fluid in humans and other animals that delivers necessary substances such as nutrients and oxygen to the cells and transports metabolic waste products away from those same cells. When it reaches the lungs, gas exchange occurs when carbon dioxide is diffused out of the blood into the pulmonary alveoli and oxygen is diffused into the blood.

This oxygenated blood is pumped to the left hand side of the heart in the pulmonary vein and enters the left atrium. From here it passes through the mitral valve, through the ventricle and taken all around the body by the aorta. Blood contains antibodies, nutrients, oxygen and much more to help the body work.

In vertebrates, it is composed of blood cells suspended in blood plasma. Plasma, which constitutes 55% of blood fluid, is mostly water (92% by volume), and contains dissipated proteins, glucose, mineral ions, hormones, carbon dioxide (plasma being the main medium for excretory product transportation), and blood cells themselves. Albumin is the main protein in plasma, and it functions to regulate the colloidal osmotic pressure of blood. The blood cells are mainly red blood cells (also called RBCs or erythrocytes), white blood cells (also called WBCs or leukocytes) and platelets. The most abundant cells in vertebrate blood are red blood cells. These contain hemoglobin, an iron-containing protein, which facilitates oxygen transport by reversibly binding to this respiratory gas and greatly increasing its solubility in blood. In contrast, carbon dioxide is almost entirely transported extracellularly dissolved in plasma as bicarbonate ion.

Vertebrate blood is bright red when its haemoglobin is oxygenated and dark red when it is deoxygenated. Some animals, such as crustaceans and mollusks, use hemocyanin to carry oxygen, instead of hemoglobin. Insects and some mollusks use a fluid called hemolymph instead of blood, the difference being that hemolymph is not contained in a closed circulatory system.

In most insects, this "blood" does not contain oxygen-carrying molecules such as hemoglobin because their bodies are small enough for their tracheal system to suffice for supplying oxygen. Jawed vertebrates have an adaptive immune system, based largely on white blood cells. White blood cells help to resist infections and parasites. Platelets are important in the clotting of blood. Arthropods, using hemolymph, have hemocytes as part of their immune system.

Blood is circulated around the body through blood vessels by the pumping action of the heart. In animals with lungs, arterial blood carries oxygen from inhaled air to the tissues of the body, and venous blood carries carbon dioxide, a waste product of metabolism produced by cells, from the tissues to the lungs to be exhaled.

Medical terms related to blood often begin with hemo- or hemato- (also spelled haemo- and haemato-) from the Greek word á7iá (haima) for "blood". In terms of anatomy and histology, blood is considered a specialized form of connective tissue, given its origin in the bones and the presence of potential molecular fibers in the form of fibrinogen.

HISTORY

According to the Oxford English Dictionary, the word "blood" dates to the oldest English, circa 1000 AD. The word is derived from Middle English, which is derived from the Old English word blôd, which is akin to the Old High German word bluot, meaning blood. The modern German word is (das) Blut.

CLASSICAL GREEK MEDICINE

Fåhræus (a Swedish physician who devised the erythrocyte sedimentation rate) suggested that the Ancient Greek system of humorism, wherein the body was thought to contain four distinct bodily fluids (associated with different temperaments), were based upon the observation of blood clotting in a transparent container.

When blood is drawn in a glass container and left undisturbed for about an hour, four different layers can be seen. A dark clot forms at the bottom (the "black bile"). Above the clot is a layer of red blood cells (the "blood"). Above this is a whitish layer of white blood cells (the "phlegm"). The top layer is clear yellow serum (the "yellow bile").

HUMAN BLOOD

The ABO blood group system was discovered in the year 1900 by Karl Landsteiner. Jan Janský is credited with the first classification of blood into the four types (A, B, AB, O) in 1907, which remains in use today. In 1907 the first blood transfusion was performed that used the ABO system to predict compatibility.

The first non-direct transfusion was performed on March 27, 1914. The Rhesus factor was discovered in 1937.

FUNCTIONS

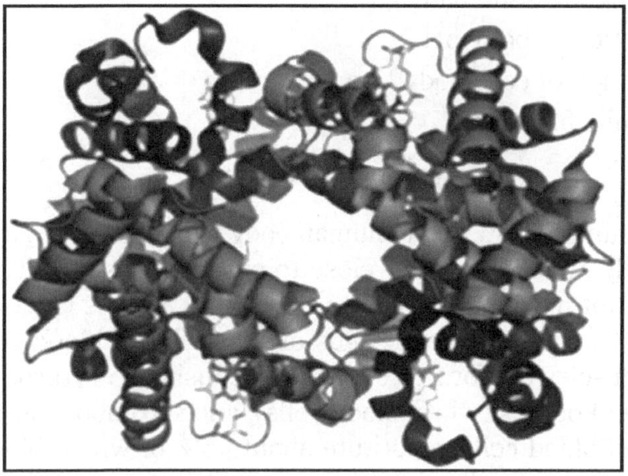

Fig. Haemoglobin, a Globular Protein Green.

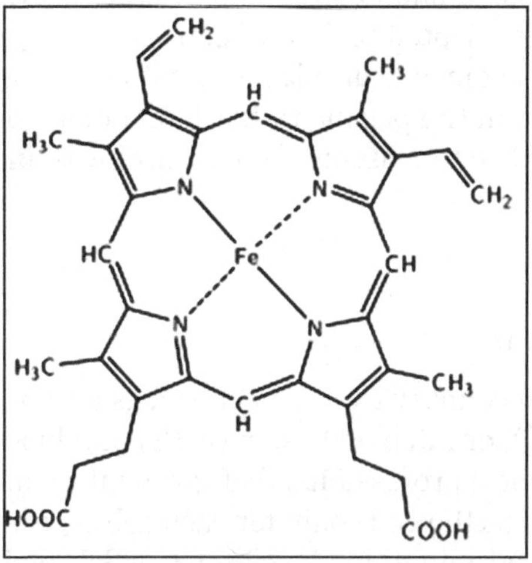

Fig. Heme.

Blood performs many important functions within the body including:

- Supply of oxygen to tissues (bound to hemoglobin, which is carried in red cells)
- Supply of nutrients such as glucose, amino acids, and fatty acids (dissolved in the blood or bound to plasma proteins (e.g., blood lipids))
- Removal of waste such as carbon dioxide, urea, and lactic acid
- Immunological functions, including circulation of white blood cells, and detection of foreign material by antibodies
- Coagulation, the response to a broken blood vessel, the conversion of blood from a liquid to a semi-solid gel to stop bleeding.

- Messenger functions, including the transport of hormones and the signaling of tissue damage
- Regulation of body pH
- Regulation of core body temperature
- Hydraulic functions

CONSTITUENTS

Blood accounts for 7% of the human body weight, with an average density of approximately 1060 kg/m, very close to pure water's density of 1000 kg/m. The average adult has a blood volume of roughly 5 litres (11 US pt), which is composed of plasma and several kinds of cells. These blood cells (which are also called corpuscles or "formed elements") consist of erythrocytes (red blood cells, RBCs), leukocytes (white blood cells), and thrombocytes (platelets). By volume, the red blood cells constitute about 45% of whole blood, the plasma about 54.3%, and white cells about 0.7%.

Whole blood (plasma and cells) exhibits non-Newtonian fluid dynamics; its flow properties are adapted to flow effectively through tiny capillary blood vessels with less resistance than plasma by itself. In addition, if all human hemoglobin were free in the plasma rather than being contained in RBCs, the circulatory fluid would be too viscous for the cardiovascular system to function effectively.

CELLS

Complete Blood Count

A complete blood count (CBC), also known as a complete blood cell count, full blood count (FBC), or full blood exam (FBE), is a blood panel requested by a doctor or other medical professional that gives information about the cells in a patient's blood, such as the cell count for each cell type and the concentrations of various proteins and minerals. A scientist or lab technician performs the requested testing and provides the requesting medical professional with the results of the CBC.

Blood counts of various types have been used for clinical purposes since the 19th century. Automated equipment to carry out complete blood counts was developed in the 1950s and 1960s.

The cells that circulate in the bloodstream are generally divided into three types: white blood cells (leukocytes), red blood cells (erythrocytes), and platelets (thrombocytes). Abnormally high or low counts may indicate the presence of many forms of disease, and hence blood counts are amongst the most commonly performed blood tests in medicine, as they can provide an overview of a patient's general health status. A CBC is routinely performed during annual physical examinations in some jurisdictions.

Medical Uses

Complete blood counts are done to monitor overall health, to screen for some diseases, to confirm a diagnosis of some medical conditions, to monitor a medical condition, and to monitor changes in the body caused by medical treatments.

For patients who need blood transfusion, a blood count may be used to get data which would help plan an amount of treatment. In such cases, the person should have only one blood count for the day, and the transfusion of red blood cells or platelets should be planned based on that. Multiple blood draws and counts throughout the day are an excessive use of phlebotomy and can lead to unnecessary additional transfusions, and the extra unnecessary treatment would be outside of medical guidelines.

Procedure

A phlebotomist collects the sample through venipuncture, drawing the blood into a test tube containing an anticoagulant (EDTA, sometimes citrate) to stop it from clotting. The sample is then transported to a laboratory. Sometimes the sample is drawn off a finger prick using a Pasteur pipette for immediate processing by an automated counter.

In the past, counting the cells in a patient's blood was performed manually, by viewing a slide prepared with a sample of the patient's blood (a blood film, or peripheral smear) under a microscope. Presently, this process is generally automated by use of an automated analyzer, with only approximately 10–20% of samples now being examined manually.

Automated Blood Count

All blood counts include a CBC count and leukocyte differential count (LDC), but modern analyzers provide much more in addition.

The blood is well mixed (though not shaken) and placed on a rack in the analyzer. This instrument has flow cells, photometers and apertures that analyze different elements in the blood. The cell counting component counts the numbers and types of different cells within the blood. The results are printed out or sent to a computer for review.

Blood counting machines aspirate a very small amount of the specimen through narrow tubing followed by an aperture and a laser flow cell. Laser eye sensors count the number of cells passing through the aperature, and can identify them; this is flow cytometry. The two main sensors used are light detectors and electrical impedance. The instrument measures the type of blood cell by analyzing data about the size and aspects of light as they pass through the cells (called front and side scatter). Other instruments measure different characteristics of the cells to categorize them. Because an automated cell counter samples and counts so many cells, the results are very precise.

However, certain abnormal cells in the blood may not be identified correctly, requiring manual review of the instrument's results and identification of any abnormal cells the instrument could not categorize.

In addition to counting, measuring and analyzing red blood cells, white blood cells and platelets, automated hematology analyzers also measure the amount of hemoglobin in the blood and within each red blood cell. This is done by adding a diluent that lyses the cells which is then pumped into a spectro-photometric measuring cuvette. The change in color of the lysate equates to the hemoglobin content of the blood. This information can be very helpful to a physician who, for example, is trying to identify the cause of a patient's anemia. If the red cells are smaller or larger than normal, or if there is a lot of variation in the size of the red cells, this data can help guide the direction of further testing and expedite the diagnostic process so patients can get the treatment they need quickly.

Manual Blood Count

Hemocytometers (counting chambers that hold a specified volume of diluted blood and divide it with grid lines) are used to calculate the number of red and white cells per litre of blood. (The dilution and grid lines are needed because there are far too many cells without those aids.)

To identify the numbers of different white cells, a blood film is made, and a large number of white blood cells (at least 100) are counted. This gives the percentage of cells that are of each type. By multiplying the percentage with the total number of white blood cells, the absolute number of each type of white cell can be obtained. Manual counting is useful in cases where automated analyzers cannot reliably count abnormal cells, such as those cells that are not present in normal patients and are only seen in peripheral blood with certain haematological conditions. Manual counting is subject to sampling error because so few cells are counted compared with automated analysis.

Medical technologists examine blood film via a microscope for some CBCs, not only to find abnormal white cells, but also because variation in the shape of red cells is an important diagnostic tool. Although automated analysers give fast, reliable results regarding the number, average size, and variation in size of red blood cells, they do not detect cells' shapes. Also, some normal patients' platelets will clump in EDTA anticoagulated blood, which causes automatic analyses to give a falsely low platelet count. The person viewing the slide in these cases will see clumps of platelets and can estimate if there are low, normal, or high numbers of platelets.

Included Tests

A complete blood count will normally include:

White cells:

- Total white blood cells: All the white cell types are given as a percentage and as an absolute number per litre. A high WBC may indicate an infection.

Red cells:

- Total red blood cells: The number of red cells is given as an absolute number per litre. Iron deficiency Anemia shows up as a Low RBC count.
  - LOINC codes for Whole Blood RBC Counting

Hemoglobin:

- Hemoglobin: The amount of hemoglobin in the blood, expressed in grams per decilitre. A low level of Hemoglobin is a sign of anemia.
  - LOINC codes for this test(Note: not all the codes in the search result are suitable for this test.)

Hematocrit:

- Hematocrit or packed cell volume (PCV): This is the fraction of whole blood volume that consists of red blood cells.
- Red blood cell indices

MCV:

- Mean corpuscular volume (MCV): the average volume of the red cells, measured in femtolitres. Anemia is classified as microcytic or macrocytic if the MCV value is above or below the expected normal range; anemias are classified as normocytic if the MCV is within the expected range. Other conditions that can affect MCV include thalassemia, reticulocytosis, alcoholism, chemotherapy, Vitamin B12 deficiency, and/or Folic acid deficiency.

MCH:

- Mean corpuscular hemoglobin (MCH): the average amount of hemoglobin per red blood cell, in picograms.

  MCHC:
- Mean corpuscular hemoglobin concentration (MCHC): the average concentration of hemoglobin in the cells.

RDW:

- Red blood cell distribution width (RDW): the variation in cellular volume of the RBC population.

A complete blood count with differential/platelet will also include:

- Neutrophil granulocytes: May indicate bacterial infection. May also be raised in acute viral infections. Because of the segmented appearance of the nucleus, neutrophils are sometimes referred to as "segs". The nucleus of less mature neutrophils is not segmented, but has a band or rod-like shape. Less mature neutrophils—those that have recently been released from the bone marrow into the bloodstream—are known as "bands" or "stabs". Stab is a German term for rod.

- Lymphocytes: Higher with some viral infections such as glandular fever and. Also raised in chronic lymphocytic leukemia (CLL). Can be decreased by HIV infection. In adults, lymphocytes are the second most common WBC type after neutrophils. In young children under age 8, lymphocytes are more common than neutrophils.
- Monocytes: May be raised in bacterial infection, tuberculosis, malaria, Rocky Mountain spotted fever, monocytic leukemia, chronic ulcerative colitis and regional enteritis
- Eosinophil granulocytes: Increased in parasitic infections, asthma, or allergic reaction.
- Basophil granulocytes: May be increased in bone marrow related conditions such as leukemia or lymphoma.

A manual count will also give information about other cells that are not normally present in peripheral blood, but may be released in certain disease processes.

Platelets:

- Platelet numbers are given, as well as information about their size and the range of sizes in the blood.
- Mean platelet volume (MPV): a measurement of the average size of platelets.

PLASMA

Blood plasma is the red liquid component of blood that normally holds the blood cells in whole blood in suspension; this makes plasma the extracellular matrix of blood cells. It makes up about 55% of the body's total blood volume. It is the intravascular fluid part of extracellular fluid (all body fluid outside of cells).

It is mostly water (up to 95% by volume), and contains dissolved proteins (6–8%) (i.e.—serum albumins, globulins, and fibrinogen), glucose, clotting factors, electrolytes (Na, Ca, Mg, HCO_3, Cl, etc.), hormones, and carbon dioxide (plasma being the main medium for excretory product transportation). Plasma also serves as the protein reserve of the human body. It plays a vital role in an intravascular osmotic effect that keeps electrolytes in balanced form and protects the body from infection and other blood disorders.

Blood plasma is prepared by spinning a tube of fresh blood containing an anticoagulant in a centrifuge until the blood cells fall to the bottom of the tube. The blood plasma is then poured or drawn off. Blood plasma has a density of approximately 1025 kg/m, or 1.025 g/ml. Blood serum is blood plasma without clotting factors; in other words, "pure" blood. Plasmapheresis is a medical therapy that involves blood plasma extraction, treatment, and reintegration. IonNormal concentration range (mmol.L)Sodium135–145Potassium3.6–5.1Chloride95–105Calcium2.1–2.8

Volume

Blood plasma volume may be expanded by or drained to extravascular fluid when there are changes in Starling forces across capillary walls. For example, when blood pressure drops in circulatory shock, Starling forces drive fluid into the interstitium, causing "third spacing."

Standing still for a prolonged period will cause an increase in transcapillary hydrostatic pressure. As a result, approximately 12% of blood plasma volume will cross into the extravascular compartment. This causes an increase in hematocrit, serum total protein, blood viscosity and, as a result of increased concentration of coagulation factors, it causes orthostatic hypercoagulability.

Medical History

The use of blood plasma as a substitute for whole blood and for transfusion purposes was proposed in March 1918, in the correspondence columns of the British Medical Journal, by Gordon R. Ward. "Dried plasmas" in powder or strips of material format were developed and first used in World War II. Prior to the United States' involvement in the war, liquid plasma and whole blood were used. The "Blood for Britain" program during the early 1940s was quite successful (and popular in the United States) based on Charles Drew's contribution. A large project began in August 1940 to collect blood in New York City hospitals for the export of plasma to Britain. Drew was appointed medical supervisor of the "Plasma for Britain" project. His notable contribution at this time was to transform the test tube methods of many blood researchers into the first successful mass production techniques.

Nonetheless, the decision was made to develop a dried plasma package for the armed forces as it would reduce breakage and make the transportation, packaging, and storage much simpler. The resulting dried plasma package came in two tin cans containing 400 cc bottles. One bottle contained enough distilled water to reconstitute the dried plasma contained within the other bottle. In about three minutes, the plasma would be ready to use and could stay fresh for around four hours.

Following the "Plasma for Britain" invention, Drew was named director of the Red Cross blood bank and assistant director of the National Research Council, in charge of blood collection for the United States Army and Navy. Drew argued against the armed forces directive that blood/plasma was to be separated by the race of the donor. Drew argued that there was no racial difference in human blood and that the policy would lead to needless deaths as soldiers and sailors were required to wait for "same race" blood.

By the end of the war the American Red Cross had provided enough blood for over six million plasma packages. Most of the surplus plasma was returned to the United States for civilian use. Serum albumin replaced dried plasma for combat use during the Korean War.

Plasma Donation

Plasma as a blood product prepared from blood donations is used in blood transfusions, typically as fresh frozen plasma (FFP) or Plasma Frozen Within 24 Hours After Phlebotomy (PF24). When donating whole blood or packed red blood cell (PRBC) transfusions, ABO blood type O- is the most desirable and is considered a "universal donor," since it has neither A nor B antigens and can be safely transfused to most recipients. Type AB+ is the "universal recipient" type for PRBC donations.

However, for plasma the situation is somewhat reversed. Blood donation centers will sometimes collect only plasma from AB donors through apheresis, as their plasma does not contain the antibodies that may cross react with recipient antigens. As such, AB is often considered the "universal donor" for plasma. Special programs exist just to cater to the male AB plasma donor, because of concerns about transfusion related acute lung injury (TRALI) and female donors who may have higher leukocyte antibodies. However, some studies show an increased risk of TRALI despite increased leukocyte antibodies in women who have been pregnant.

Following fears of vCJD being spread through the blood supply, the British government began to import American plasma to meet its needs. In 2002, it purchased Life Resources Incorporated in order to be able to secure plasma supply from the United States.

The company became Plasma Resources UK (PRUK). In 2013, the British government sold an 80% stake in PRUK to American hedge fund Bain Capital, in a deal estimated to be worth £230 million. The sale was met with criticism in the UK, over both the oversight of the company, and the impact on British blood needs. Confusion arose as some thought that the blood of British blood donors voluntarily donating to NHS Blood and Transplant Services would be sold for profit. Currently, plasma donated in Britain is still not used due to concerns about vCJD, and is incinerated after component parts have been separated from it.

Safety and Compensation to Donors

In contrast to whole blood donation, plasma donation may be done much more often (2 times per week in some jurisdictions, compared to once per 8 weeks for whole blood donation), because the whole blood cells are returned to the body during the donation process. Most plasma is naturally replaced by the body within 48 hours. In some jurisdictions, and in contrast to whole blood donation, plasma donors may be compensated financially ($10–$100 per week for 3–7 hours of time, normally). It is usually defined as a compensation for time and trouble rather than a "purchase" or "sale" of a body part, for various legal, ethical, and cultural reasons.

Synthetic Blood Plasma

Simulated body fluid (SBF) is a solution having a similar ion concentration to that of human blood plasma. SBF is normally used for the surface modification of metallic implants, and more recently in gene delivery application.

ACID–BASE HOMEOSTASIS

Acid–base homeostasis is the part of human homeostasis concerning the proper balance between acids and bases, also called body pH. The body is very sensitive to its pH level, so strong mechanisms exist to maintain it. Outside the acceptable range of pH, proteins are denatured and digested, enzymes lose their ability to function, and death may occur.

Mechanism

The body's acid–base balance is normally tightly regulated by buffering agents, the respiratory system, and the renal system, keeping the arterial blood pH between 7.38 and 7.42. Several buffering agents that reversibly bind hydrogen ions and impede any change in pH exist. Extracellular buffers include bicarbonate and ammonia, whereas proteins and phosphate act as intracellular buffers; the relationship between multiple buffers in the same solution is described by the isohydric principle. The bicarbonate buffering system is especially key, as carbon dioxide (CO_2) can be shifted through carbonic acid (H_2CO_3) to hydrogen ions and bicarbonate ($HCO\text{-}_3$) as shown below.

$$H_2O + CO_2 \leftrightarrow H_2CO_3 \leftrightarrow H^+ + HCO_3^-$$

Acid–base imbalances that overcome the buffer system can be compensated in the short term by changing the rate of ventilation. This alters the concentration of carbon dioxide in the blood, shifting the above reaction according to Le Chatelier's principle, which in turn alters the pH. For instance, if the blood pH drops too low (acidemia), the body will compensate by increasing breathing thereby expelling CO_2, and shifting the above reaction to the left such that fewer hydrogen ions are free; thus the pH will rise back to normal. For alkalemia, the opposite occurs.

The kidneys are slower to compensate, but renal physiology has several powerful mechanisms to control pH by the excretion of excess acid or base. In response to acidosis, tubular cells reabsorb more bicarbonate from the tubular fluid, collecting duct cells secrete more hydrogen and generate more bicarbonate, and ammoniagenesis leads to increased formation of the NH3 buffer. In responses to alkalosis, the kidney may excrete more bicarbonate by decreasing hydrogen ion secretion from the tubular epithelial cells, and lowering rates of glutamine metabolism and ammonium excretion.

Imbalance

Acid–base imbalance occurs when a significant insult causes the blood pH to shift out of the normal range (7.35 to 7.45). In the fetus, the normal range differs based on which umbilical vessel is sampled (umbilical vein pH is normally 7.25 to 7.45; umbilical artery pH is normally 7.18 to 7.38). An excess of acid in the blood is called acidemia and an excess of base is called alkalemia. The process that causes the imbalance is classified based on the etiology of the disturbance (respiratory or metabolic) and the direction of change in pH (acidosis or alkalosis).

There are four basic processes: metabolic acidosis, respiratory acidosis, metabolic alkalosis, and respiratory alkalosis. One or a combination may occur at any given time.

BLOOD IN NON-MAMMALIAN VERTEBRATES

Human blood is typical of that of mammals, although the precise details concerning cell numbers, size, protein structure, and so on, vary somewhat between species.

In non-mammalian vertebrates, however, there are some key differences:

- Red blood cells of non-mammalian vertebrates are flattened and ovoid in form, and retain their cell nuclei
- There is considerable variation in the types and proportions of white blood cells; for example, acidophils are generally more common than in humans
- Platelets are unique to mammals; in other vertebrates, small nucleated, spindle cells called thrombocytes are responsible for blood clotting instead

PHYSIOLOGY

CARDIOVASCULAR SYSTEM

Blood is circulated around the body through blood vessels by the pumping action of the heart. In humans, blood is pumped from the strong left ventricle of the heart through arteries to peripheral tissues and returns to the right atrium of the heart through veins. It then enters the right ventricle and is pumped through the pulmonary artery to the lungs and returns to the left atrium through the pulmonary veins. Blood then enters the left ventricle to be circulated again. Arterial blood carries oxygen from inhaled air to all of the cells of the body, and venous blood carries carbon dioxide, a waste product of metabolism by cells, to the lungs to be exhaled. However, one exception includes pulmonary arteries, which contain the most deoxygenated blood in the body, while the pulmonary veins contain oxygenated blood.

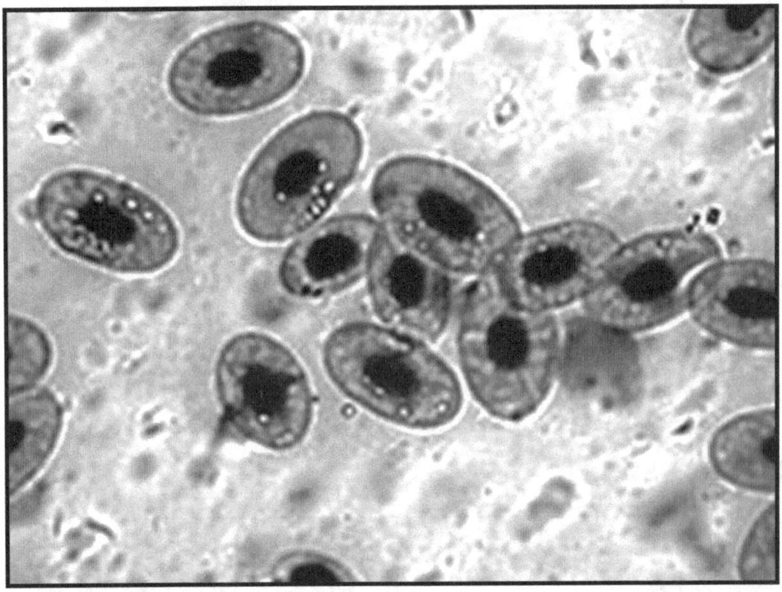

Fig. Frog Red Blood Cells Magnified 1000 Times.

Additional return flow may be generated by the movement of skeletal muscles, which can compress veins and push blood through the valves in veins toward the right atrium. The blood circulation was famously described by William Harvey in 1628.

PRODUCTION AND DEGRADATION OF BLOOD CELLS

In vertebrates, the various cells of blood are made in the bone marrow in a process called hematopoiesis, which includes erythropoiesis, the production of red blood cells; and myelopoiesis, the production of white blood cells and platelets.

During childhood, almost every human bone produces red blood cells; as adults, red blood cell production is limited to the larger bones: the bodies of the vertebrae, the breastbone (sternum), the ribcage, the pelvic bones, and the bones of the upper arms and legs. In addition, during childhood, the thymus gland, found in the mediastinum, is an important source of T lymphocytes. The proteinaceous component of blood (including clotting proteins) is produced predominantly by the liver, while hormones are produced by the endocrine glands and the watery fraction is regulated by the hypothalamus and maintained by the kidney.

Healthy erythrocytes have a plasma life of about 120 days before they are degraded by the spleen, and the Kupffer cells in the liver.

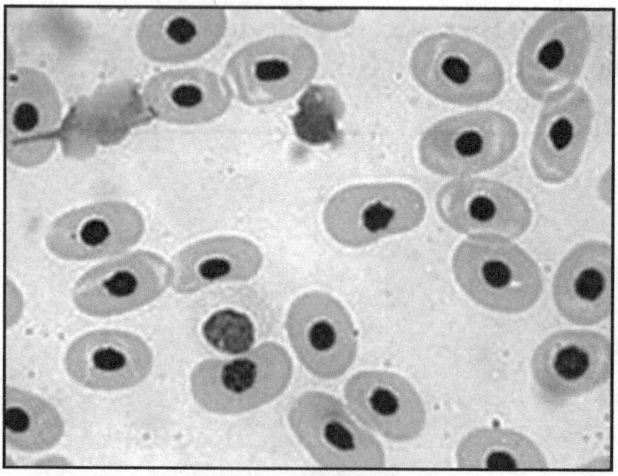

Fig. Turtle Red Blood Cells Magnified 1000 Times.

The liver also clears some proteins, lipids, and amino acids. The kidney actively secretes waste products into the urine.

OXYGEN TRANSPORT

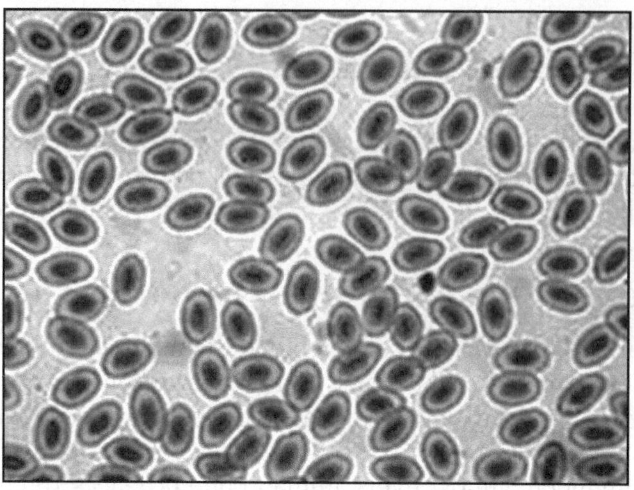

Fig. Chicken Red Blood Cells Magnified 1000 Times.

About 98.5% of the oxygen in a sample of arterial blood in a healthy human breathing air at sea-level pressure is chemically combined with the Hgb. About 1.5% is physically dissolved in the other blood liquids and not connected to Hgb. The hemoglobin molecule is the primary transporter of oxygen in mammals and many other species (for exceptions, see below). Hemoglobin has an oxygen binding capacity of between 1.36 and 1.37 ml O_2 per gram hemoglobin, which increases the total blood oxygen capacity seventyfold, compared to if oxygen solely were carried by its solubility of 0.03 ml O_2 per liter blood per mm Hg partial pressure of oxygen (approximately 100 mm Hg in arteries).

With the exception of pulmonary and umbilical arteries and their corresponding veins, arteries carry oxygenated blood away from the heart and deliver it to the body via arterioles and capillaries, where the oxygen is consumed; afterwards, venules, and veins carry deoxygenated blood back to the heart.

Under normal conditions in adult humans at rest; hemoglobin in blood leaving the lungs is about 98–99% saturated with oxygen, achieving an oxygen delivery of between 950 and 1150 ml/min to the body. In a healthy adult at rest, oxygen consumption is approximately 200 - 250 ml/min, and deoxygenated blood returning to the lungs is still approximately 75% (70 to 78%) saturated. Increased oxygen consumption during sustained exercise reduces the oxygen saturation of venous blood, which can reach less than 15% in a trained athlete; although breathing rate and blood flow increase to compensate, oxygen saturation in arterial blood can drop to 95% or less under these conditions. Oxygen saturation this low is considered dangerous in an individual at rest (for instance, during surgery under anesthesia). Sustained hypoxia (oxygenation of less than 90%), is dangerous to health, and severe hypoxia (saturations of less than 30%) may be rapidly fatal.

A fetus, receiving oxygen via the placenta, is exposed to much lower oxygen pressures (about 21% of the level found in an adult's lungs), and, so, fetuses produce another form of hemoglobin with a much higher affinity for oxygen (hemoglobin F) in order to function under these conditions.

CARBON DIOXIDE TRANSPORT

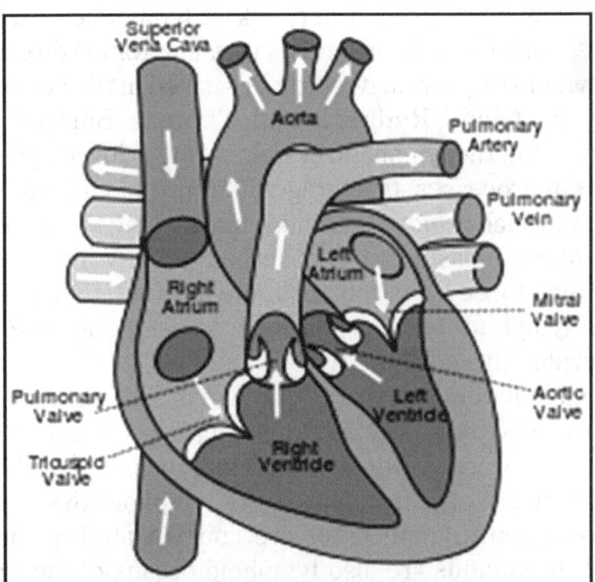

Fig. The Circulation of Blood Through the Human Heart.

CO_2 is carried in blood in three different ways. (The exact percentages vary depending whether it is arterial or venous blood). Most of it (about 70%) is converted to bicarbonate ions HCO"3 by the enzyme carbonic anhydrase in the red blood cells by the reaction CO_2 + H_2O ! H_2CO_3 ! H + HCO"3; about 7% is dissolved in the plasma; and about 23% is bound to hemoglobin as carbamino compounds. Hemoglobin, the main oxygen-carrying molecule in red blood cells, carries both oxygen and carbon dioxide. However, the CO_2 bound to hemoglobin does not bind to the same site as oxygen. Instead, it combines with the N-terminal groups on the four globin chains. However, because of allosteric effects on the hemoglobin molecule, the binding of CO_2 decreases the amount of oxygen that is bound for a given partial pressure of oxygen. The decreased binding to carbon dioxide in the blood due to increased oxygen levels is known as the Haldane effect, and is important in the transport of carbon dioxide from the tissues to the lungs. A rise in the partial pressure of CO_2 or a lower pH will cause offloading of oxygen from hemoglobin, which is known as the Bohr effect.

TRANSPORT OF HYDROGEN IONS

Some oxyhemoglobin loses oxygen and becomes deoxyhemoglobin. Deoxyhemoglobin binds most of the hydrogen ions as it has a much greater affinity for more hydrogen than does oxyhemoglobin.

LYMPHATIC SYSTEM

The lymphatic system is part of the circulatory system and a vital part of the immune system, comprising a network of lymphatic vessels that carry a clear fluid called lymph (from Latin lympha meaning water) directionally towards the heart. The lymphatic system was first described in the seventeenth century independently by Olaus Rudbeck and Thomas Bartholin. Unlike the cardiovascular system, the lymphatic system is not a closed system. The human circulatory system processes an average of 20 litres of blood per day through capillary filtration, which removes plasma while leaving the blood cells. Roughly 17 litres of the filtered plasma are reabsorbed directly into the blood vessels, while the remaining three litres remain in the interstitial fluid. One of the main functions of the lymph system is to provide an accessory return route to the blood for the surplus three litres.

The other main function is that of defense in the immune system. Lymph is very similar to blood plasma: it contains lymphocytes and other white blood cells. It also contains waste products and debris of cells together with bacteria and protein. Associated organs composed of lymphoid tissue are the sites of lymphocyte production. Lymphocytes are concentrated in the lymph nodes. The spleen and the thymus are also lymphoid organs of the immune system. The tonsils are lymphoid organs that are also associated with the digestive system. Lymphoid tissues contain lymphocytes, and also contain other types of cells for support. The system also includes all the structures dedicated to the circulation and production of lymphocytes (the primary cellular component of lymph), which also includes the bone marrow, and the lymphoid tissue associated with the digestive system.

The blood does not come into direct contact with the parenchymal cells and tissues in the body (except in case of an injury causing rupture of one or more blood vessels), but constituents of the blood first exit the microvascular exchange blood vessels to become interstitial fluid, which comes into contact with the parenchymal cells of the body. Lymph is the fluid that is formed when interstitial fluid enters the initial lymphatic vessels of the lymphatic system. The lymph is then moved along the lymphatic vessel network by either intrinsic contractions of the lymphatic passages or by extrinsic compression of the lymphatic vessels via external tissue forces (e.g., the contractions of skeletal muscles), or by lymph hearts in some animals. The organization of lymph nodes and drainage follows the organization of the body into external and internal regions; therefore, the lymphatic drainage of the head, limbs, and body cavity walls follows an external route, and the lymphatic drainage of the thorax, abdomen, and pelvic cavities follows an internal route. Eventually, the lymph vessels empty into the lymphatic ducts, which drain into one of the two subclavian veins, near their junction with the internal jugular veins.

Structure

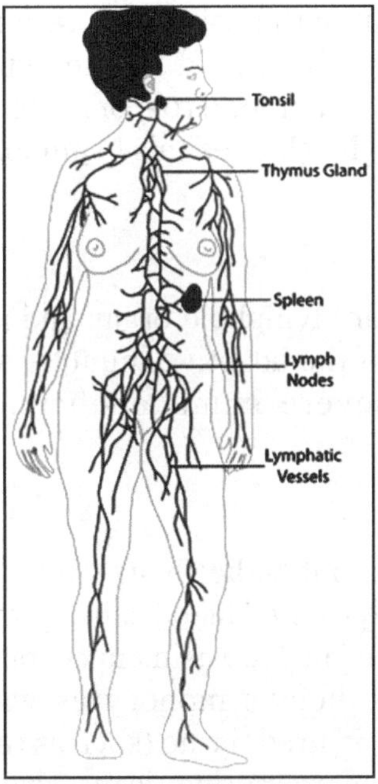

Fig. Lymphatic System.

The lymphatic system consists of lymphatic organs, a conducting network of lymphatic vessels, and the circulating lymph.

The thymus and the bone marrow constitute the primary lymphoid organs involved in the production and early clonal selection of lymphocyte tissues. Bone marrow is responsible for both the creation of T cells and the production and maturation of B cells. From the bone marrow, B cells immediately join the circulatory system and travel to secondary lymphoid organs in search of pathogens. T cells, on the other hand, travel from the bone marrow to the thymus, where they develop further. Mature T cells join B cells in search of pathogens. The other 95% of T cells begin a process of apoptosis (programmed cell death).

The central or primary lymphoid organs generate lymphocytes from immature progenitor cells. Secondary or peripheral lymphoid organs, which include lymph nodes and the spleen, maintain mature naive lymphocytes and initiate an adaptive immune response. The peripheral lymphoid organs are the sites of lymphocyte activation by antigens. Activation leads to clonal expansion and affinity maturation. Mature lymphocytes recirculate between the blood and the peripheral lymphoid organs until they encounter their specific antigen.

Secondary lymphoid tissue provides the environment for the foreign or altered native molecules (antigens) to interact with the lymphocytes. It is exemplified by the lymph nodes, and the lymphoid follicles in tonsils, Peyer's patches, spleen, adenoids, skin, etc. that are associated with the mucosa-associated lymphoid tissue (MALT). In the gastrointestinal wall the appendix has mucosa resembling that of the colon, but here it is heavily infiltrated with lymphocytes. The tertiary lymphoid tissue typically contains far fewer lymphocytes, and assumes an immune role only when challenged with antigens that result in inflammation. It achieves this by importing the lymphocytes from blood and lymph.)

Thymus

The thymus is a primary lymphoid organ and is the site of maturation for T cells, the lymphocytes of the adaptive immune system. The loss or lack of the thymus results in severe immunodeficiency and subsequent high susceptibility to infection.

Spleen

The spleen synthesizes antibodies in its white pulp and removes antibody-coated bacteria and antibody-coated blood cells by way of blood and lymph node circulation. A study published in 2009 using mice found that the spleen contains, in its reserve, half of the body's monocytes within the red pulp. These monocytes, upon moving to injured tissue (such as the heart), turn into dendritic cells and macrophages while promoting tissue healing. The spleen is a center of activity of the mononuclear phagocyte system and can be considered analogous to a large lymph node, as its absence causes a predisposition to certain infections.

Like the thymus, the spleen has only efferent lymphatic vessels. Both the short gastric arteries and the splenic artery supply it with blood. The germinal centers are supplied by arterioles called penicilliary radicles.

Up to the fifth month of prenatal development the spleen creates red blood cells. After birth the bone marrow is solely responsible for hematopoiesis. As a major lymphoid organ and a central player in the reticuloendothelial system, the spleen retains the ability to produce lymphocytes. The spleen stores red blood cells and lymphocytes. It can store enough blood cells to help in an emergency. Up to 25% of lymphocytes can be stored at any one time.

Lymph Nodes

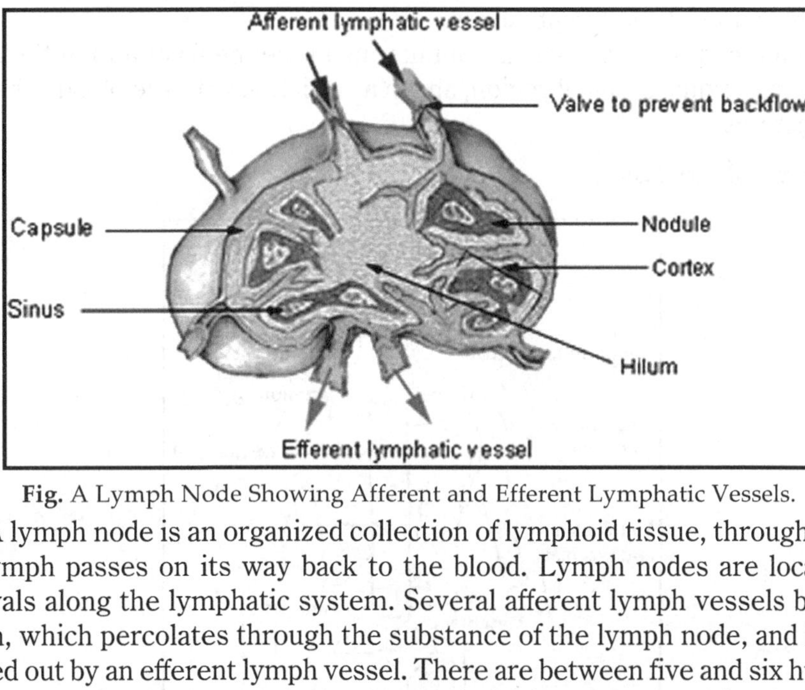

Fig. A Lymph Node Showing Afferent and Efferent Lymphatic Vessels.

A lymph node is an organized collection of lymphoid tissue, through which the lymph passes on its way back to the blood. Lymph nodes are located at intervals along the lymphatic system. Several afferent lymph vessels bring in lymph, which percolates through the substance of the lymph node, and is then drained out by an efferent lymph vessel. There are between five and six hundred lymph nodes in the human body, many of which are grouped in clusters in different regions as in the underarm and abdominal areas. Lymph node clusters are commonly found at the base of limbs (groin, armpits) and in the neck, where lymph is collected from regions of the body likely to sustain pathogen contamination from injuries.

The substance of a lymph node consists of lymphoid follicles in an outer portion called the cortex. The inner portion of the node is called the medulla, which is surrounded by the cortex on all sides except for a portion known as the hilum. The hilum presents as a depression on the surface of the lymph node, causing the otherwise spherical lymph node to be bean-shaped or ovoid. The efferent lymph vessel directly emerges from the lymph node at the hilum. The arteries and veins supplying the lymph node with blood enter and exit through the hilum.

The region of the lymph node called the paracortex immediately surrounds the medulla. Unlike the cortex, which has mostly immature T cells, or thymocytes, the paracortex has a mixture of immature and mature T cells. Lymphocytes enter the lymph nodes through specialised high endothelial venules found in the paracortex.

A lymph follicle is a dense collection of lymphocytes, the number, size and configuration of which change in accordance with the functional state of the lymph node. For example, the follicles expand significantly when encountering a foreign antigen. The selection of B cells, or B lymphocytes, occurs in the germinal center of the lymph nodes.

Lymph nodes are particularly numerous in the mediastinum in the chest, neck, pelvis, axilla, inguinal region, and in association with the blood vessels of the intestines.

Other Lymphoid Tissue

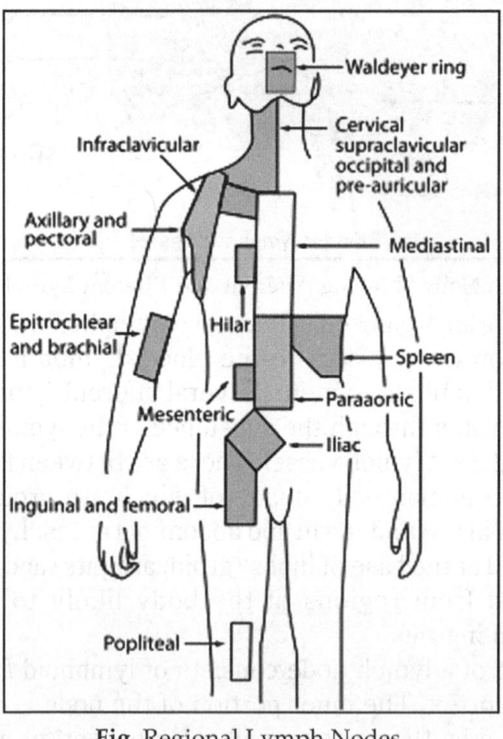

Fig. Regional Lymph Nodes.

Lymphoid tissue associated with the lymphatic system is concerned with immune functions in defending the body against infections and the spread of tumors. It consists of connective tissue formed of reticular fibers, with various types of leukocytes, (white blood cells), mostly lymphocytes enmeshed in it, through which the lymph passes. Regions of the lymphoid tissue that are densely packed with lymphocytes are known as lymphoid follicles. Lymphoid tissue can either be structurally well organized as lymph nodes or may consist of loosely organized lymphoid follicles known as the mucosa-associated lymphoid tissue.

Lymphatics

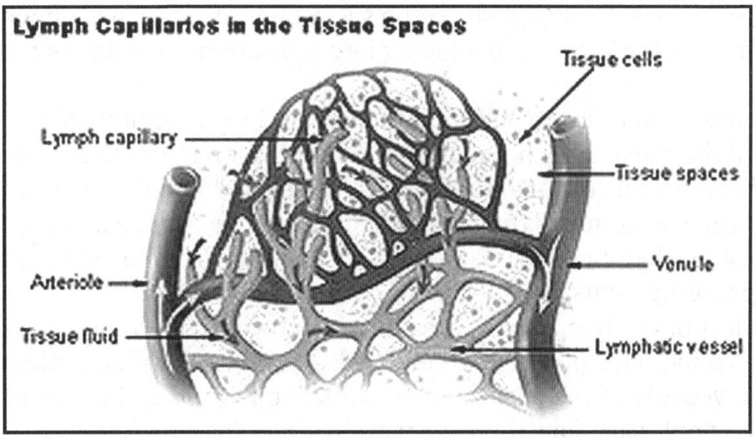

Fig. Lymph Capillaries in the Tissue Spaces.

The lymphatic vessels, also called lymph vessels, conduct lymph between different parts of the body. They include the tubular vessels of the lymph capillaries, and the larger collecting vessels–the right lymphatic duct and the thoracic duct (the left lymphatic duct). The lymph capillaries are mainly responsible for the absorption of interstitial fluid from the tissues, while lymph vessels propel the absorbed fluid forward into the larger collecting ducts, where it ultimately returns to the bloodstream via one of the subclavian veins. These vessels are also called the lymphatic channels or simply lymphatics.

The lymphatics are responsible for maintaining the balance of the body fluids. Its network of capillaries and collecting lymphatic vessels work to efficiently drain and transport extravasated fluid, along with proteins and antigens, back to the circulatory system. Numerous intraluminal valves in the vessels ensure a unidirectional flow of lymph without reflux. Two valve systems are used to achieve this one directional flow—a primary and a secondary valve system. The capillaries are blind-ended, and the valves at the ends of capillaries use specialised junctions together with anchoring filaments to allow a unidirectional flow to the primary vessels. The collecting lymphatics, however, act to propel the lymph by the combined actions of the intraluminal valves and lymphatic muscle cells.

Development

Lymphatic tissues begin to develop by the end of the fifth week of embryonic development. Lymphatic vessels develop from lymph sacs that arise from developing veins, which are derived from mesoderm.

The first lymph sacs to appear are the paired jugular lymph sacs at the junction of the internal jugular and subclavian veins. From the jugular lymph sacs, lymphatic capillary plexuses spread to the thorax, upper limbs, neck and head. Some of the plexuses enlarge and form lymphatic vessels in their respective regions. Each jugular lymph sac retains at least one connection with its jugular vein, the left one developing into the superior portion of the thoracic duct.

The next lymph sac to appear is the unpaired retroperitoneal lymph sac at the root of the mesentery of the intestine. It develops from the primitive vena cava and mesonephric veins. Capillary plexuses and lymphatic vessels spread from the retroperitoneal lymph sac to the abdominal viscera and diaphragm. The sac establishes connections with the cisterna chyli but loses its connections with neighboring veins.

The last of the lymph sacs, the paired posterior lymph sacs, develop from the iliac veins. The posterior lymph sacs produce capillary plexuses and lymphatic vessels of the abdominal wall, pelvic region, and lower limbs. The posterior lymph sacs join the cisterna chyli and lose their connections with adjacent veins.

With the exception of the anterior part of the sac from which the cisterna chyli develops, all lymph sacs become invaded by mesenchymal cells and are converted into groups of lymph nodes.

The spleen develops from mesenchymal cells between layers of the dorsal mesentery of the stomach. The thymus arises as an outgrowth of the third pharyngeal pouch.

Function

The lymphatic system has multiple interrelated functions:

- It is responsible for the removal of interstitial fluid from tissues
- It absorbs and transports fatty acids and fats as chyle from the digestive system
- It transports white blood cells to and from the lymph nodes into the bones
- The lymph transports antigen-presenting cells (APCs), such as dendritic cells, to the lymph nodes where an immune response is stimulated.

Function of the Fatty Acid Transport System

Lymph vessels called lacteals are present in the lining of the gastrointestinal tract, predominantly in the small intestine. While most other nutrients absorbed by the small intestine are passed on to the portal venous system to drain via the portal vein into the liver for processing, fats (lipids) are passed on to the lymphatic system to be transported to the blood circulation via the thoracic duct. (There are exceptions, for example medium-chain triglycerides (MCTs) are fatty acid esters of glycerol that passively diffuse from the GI tract to the portal system.) The enriched lymph originating in the lymphatics of the small intestine is called chyle. The nutrients that are released to the circulatory system are processed by the liver, having passed through the systemic circulation.

Clinical Significance

The study of lymphatic drainage of various organs is important in diagnosis, prognosis, and treatment of cancer. The lymphatic system, because of its physical proximity to many tissues of the body, is responsible for carrying cancerous cells between the various parts of the body in a process called metastasis. The intervening lymph nodes can trap the cancer cells. If they are not successful in destroying the cancer cells the nodes may become sites of secondary tumors.

Lymphadenopathy

Lymphadenopathy refers to one or more enlarged lymph nodes. Small groups or individually enlarged lymph nodes are generally reactive in response to infection or inflammation. This is called local lymphadenopathy. When many lymph nodes in different areas of the body are involved, this is called generalised lymphadenopathy. Generalised lymphadenopathy may be caused by infections such as infectious mononucleosis, tuberculosis and HIV, connective tissue diseases such as SLE and rheumatoid arthritis, and cancers, including both cancers of tissue within lymph nodes, discussed below, and metastasis of cancerous cells from other parts of the body, that have arrived via the lymphatic system.

Lymphedema

Lymphedema is the swelling caused by the accumulation of lymph, which may occur if the lymphatic system is damaged or has malformations. It usually affects limbs, though the face, neck and abdomen may also be affected. In an extreme state, called elephantiasis, the edema progresses to the extent that the skin becomes thick with an appearance similar to the skin on elephant limbs.

Causes are unknown in most cases, but sometimes there is a previous history of severe infection, usually caused by a parasitic disease, such as lymphatic filariasis. Lymphangiomatosis is a disease involving multiple cysts or lesions formed from lymphatic vessels. Lymphedema can also occur after surgical removal of cancerous lymph nodes in the armpit (causing the arm to swell due to poor lymphatic drainage) or groin (causing swelling of the leg). Treatment is by massage, and is not permanent.

Cancer

Cancer of the lymphatic system can be primary or secondary. Lymphoma refers to cancer that arises from lymphatic tissue. Lymphoid leukemias and lymphomas are now considered to be tumors of the same type of cell lineage. They are called "leukemia" when in the blood or marrow and "lymphoma" when in lymphatic tissue. They are grouped together under the name "lymphoid malignancy".

Lymphoma is generally considered as either Hodgkin lymphoma or non-Hodgkin lymphoma. Hodgkin lymphoma is characterised by a particular type of cell, called a Reed–Sternberg cell, visible under microscope. It is associated with past infection with the Epstein-Barr Virus, and generally causes a painless "rubbery" lymphadenopathy. It is staged, using Ann Arbor staging. Chemotherapy generally involves the ABVD and may also involved radiotherapy. Non-Hodgkin lymphoma is a cancer characterised by increased proliferation of B-cells or T-cells, generally occurs in an older age group than Hodgkin lymphoma. It is treated according to whether it is high-grade or low-grade, and carries a poorer prognosis than Hodgkin lymphoma.

Lymphangiosarcoma is a malignant soft tissue tumor, whereas lymphangioma is a benign tumor occurring frequently in association with Turner syndrome. Lymphangioleiomyomatosis is a benign tumor of the smooth muscles of the lymphatics that occurs in the lungs.

History

Hippocrates, in 5th century BC, was one of the first people to mention the lymphatic system. In his work On Joints, he briefly mentioned the lymph nodes in one sentence. Rufus of Ephesus, a Roman physician, identified the axillary, inguinal and mesenteric lymph nodes as well as the thymus during the 1st to 2nd century AD.

The first mention of lymphatic vessels was in 3rd century BC by Herophilos, a Greek anatomist living in Alexandria, who incorrectly concluded that the "absorptive veins of the lymphatics," by which he meant the lacteals (lymph vessels of the intestines), drained into the hepatic portal veins, and thus into the liver. The findings of Ruphus and Herophilos were further propagated by the Greek physician Galen, who described the lacteals and mesenteric lymph nodes which he observed in his dissection of apes and pigs in the 2nd century AD.

In the mid 16th century, Gabriele Falloppio (discoverer of the fallopian tubes), described what are now known as the lacteals as "coursing over the intestines full of yellow matter." In about 1563 Bartolomeo Eustachi, a professor of anatomy, described the thoracic duct in horses as vena alba thoracis. The next breakthrough came when in 1622 a physician, Gaspare Aselli, identified lymphatic vessels of the intestines in dogs and termed them venae alba et lacteae, which is now known as simply the lacteals.

The lacteals were termed the fourth kind of vessels (the other three being the artery, vein and nerve, which was then believed to be a type of vessel), and disproved Galen's assertion that chyle was carried by the veins. But, he still believed that the lacteals carried the chyle to the liver (as taught by Galen). He also identified the thoracic duct but failed to notice its connection with the lacteals.

This connection was established by Jean Pecquet in 1651, who found a white fluid mixing with blood in a dog's heart. He suspected that fluid to be chyle as its flow increased when abdominal pressure was applied. He traced this fluid to the thoracic duct, which he then followed to a chyle-filled sac he called the chyli receptaculum, which is now known as the cisternae chyli; further investigations led him to find that lacteals' contents enter the venous system via the thoracic duct. Thus, it was proven convincingly that the lacteals did not terminate in the liver, thus disproving Galen's second idea: that the chyle flowed to the liver. Johann Veslingius drew the earliest sketches of the lacteals in humans in 1647.

The idea that blood recirculates through the body rather than being produced anew by the liver and the heart was first accepted as a result of works of William Harvey—a work he published in 1628. In 1652, Olaus Rudbeck (1630–1702), a Swede, discovered certain transparent vessels in the liver that contained clear fluid (and not white), and thus named them hepatico-aqueous vessels. He also learned that they emptied into the thoracic duct, and that they had valves. He announced his findings in the court of Queen Christina of Sweden, but did not publish his findings for a year, and in the interim similar findings were published by Thomas Bartholin, who additionally published that such vessels are present everywhere in the body, not just in the liver. He is also the one to have named them "lymphatic vessels." This had resulted in a bitter dispute between one of Bartholin's pupils, Martin Bogdan, and Rudbeck, whom he accused of plagiarism.

Galen's ideas prevailed in medicine until the 17th century. It was believed that blood was produced by the liver from chyle contaminated with ailments by the intestine and stomach, to which various spirits were added by other organs, and that this blood was consumed by all the organs of the body. This theory required that the blood be consumed and produced many times over. Even in the 17th century, his ideas were defended by some physicians.

THERMOREGULATION

Blood circulation transports heat throughout the body, and adjustments to this flow are an important part of thermoregulation. Increasing blood flow to the surface (e.g., during warm weather or strenuous exercise) causes warmer skin, resulting in faster heat loss. In contrast, when the external temperature is low, blood flow to the extremities and surface of the skin is reduced and to prevent heat loss and is circulated to the important organs of the body, preferentially.

HYDRAULIC FUNCTIONS

The restriction of blood flow can also be used in specialized tissues to cause engorgement, resulting in an erection of that tissue; examples are the erectile tissue in the penis and clitoris.

Another example of a hydraulic function is the jumping spider, in which blood forced into the legs under pressure causes them to straighten for a powerful jump, without the need for bulky muscular legs.

INVERTEBRATES

In insects, the blood (more properly called hemolymph) is not involved in the transport of oxygen. (Openings called tracheae allow oxygen from the air to diffuse directly to the tissues). Insect blood moves nutrients to the tissues and removes waste products in an open system.

Other invertebrates use respiratory proteins to increase the oxygen-carrying capacity. Hemoglobin is the most common respiratory protein found in nature.

Hemocyanin (blue) contains copper and is found in crustaceans and mollusks. It is thought that tunicates (sea squirts) might use vanabins (proteins containing vanadium) for respiratory pigment (bright-green, blue, or orange).

In many invertebrates, these oxygen-carrying proteins are freely soluble in the blood; in vertebrates they are contained in specialized red blood cells, allowing for a higher concentration of respiratory pigments without increasing viscosity or damaging blood filtering organs like the kidneys.

Giant tube worms have unusual hemoglobins that allow them to live in extraordinary environments. These hemoglobins also carry sulfides normally fatal in other animals.

COLOR

The coloring matter of blood (hemochrome) is largely due to the protein in the blood responsible for oxygen transport. Different groups of organisms use different proteins.

HEMOGLOBIN

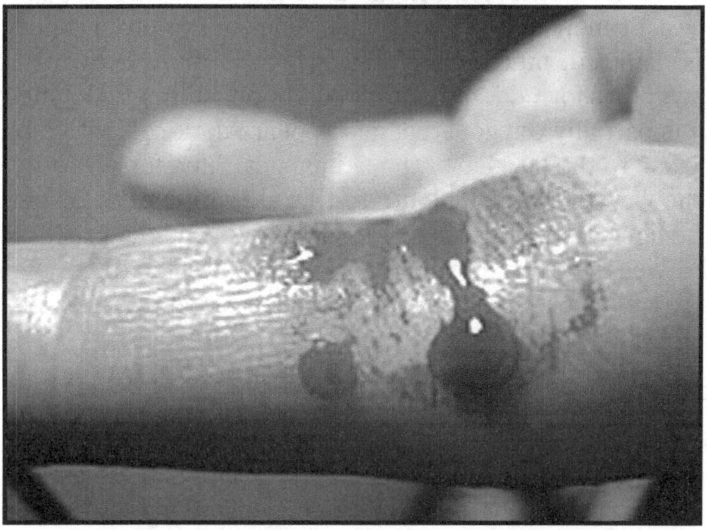

Fig. Capillary Blood from a Bleeding Finger.

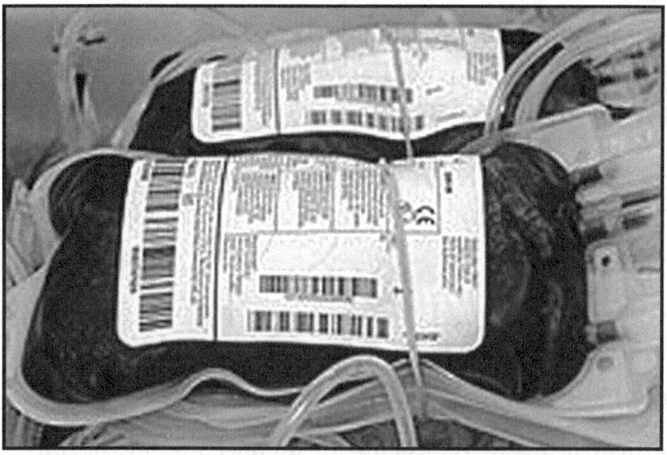

Fig. Venous Blood Collected During Blood Donation.

Hemoglobin is the principal determinant of the color of blood in vertebrates. Each molecule has four heme groups, and their interaction with various molecules alters the exact color. In vertebrates and other hemoglobin-using creatures, arterial blood and capillary blood are bright red, as oxygen imparts a strong red color to the heme group. Deoxygenated blood is a darker shade of red; this is present in veins, and can be seen during blood donation and when venous blood samples are taken. This is because the spectrum of light absorbed by hemoglobin differs between the oxygenated and deoxygenated states.

Blood in carbon monoxide poisoning is bright red, because carbon monoxide causes the formation of carboxyhemoglobin. In cyanide poisoning, the body cannot utilize oxygen, so the venous blood remains oxygenated, increasing the redness. There are some conditions affecting the heme groups present in hemoglobin that can make the skin appear blue—a symptom called cyanosis. If the heme is oxidized, methaemoglobin, which is more brownish and cannot transport oxygen, is formed. In the rare condition sulfhemoglobinemia, arterial hemoglobin is partially oxygenated, and appears dark red with a bluish hue.

Veins close to the surface of the skin appear blue for a variety of reasons. However, the factors that contribute to this alteration of color perception are related to the light-scattering properties of the skin and the processing of visual input by the visual cortex, rather than the actual color of the venous blood. Skinks in the genus Prasinohaema have green blood due to a buildup of the waste product biliverdin.

Hemocyanin

The blood of most mollusks – including cephalopods and gastropods – as well as some arthropods, such as horseshoe crabs, is blue, as it contains the copper-containing protein hemocyanin at concentrations of about 50 grams per liter. Hemocyanin is colorless when deoxygenated and dark blue when oxygenated.

The blood in the circulation of these creatures, which generally live in cold environments with low oxygen tensions, is grey-white to pale yellow, and it turns dark blue when exposed to the oxygen in the air, as seen when they bleed. This is due to change in color of hemocyanin when it is oxidized. Hemocyanin carries oxygen in extracellular fluid, which is in contrast to the intracellular oxygen transport in mammals by hemoglobin in RBCs.

Chlorocruorin

The blood of most annelid worms and some marine polychaetes use chlorocruorin to transport oxygen. It is green in color in dilute solutions.

Hemerythrin

Hemerythrin is used for oxygen transport in the marine invertebrates sipunculids, priapulids, brachiopods, and the annelid worm, magelona. Hemerythrin is violet-pink when oxygenated.

Hemovanadin

The blood of some species of ascidians and tunicates, also known as sea squirts, contains proteins called vanabins. These proteins are based on vanadium, and give the creatures a concentration of vanadium in their bodies 100 times higher than the surrounding sea water. Unlike hemocyanin and hemoglobin, hemovanadin is not an oxygen carrier. When exposed to oxygen, however, vanabins turn a mustard yellow.

PATHOLOGY

GENERAL MEDICAL DISORDERS

- Disorders of volume
  - Injury can cause blood loss through bleeding. A healthy adult can lose almost 20% of blood volume (1 L) before the first symptom, restlessness, begins, and 40% of volume (2 L) before shock sets in. Thrombocytes are important for blood coagulation and the formation of blood clots, which can stop bleeding. Trauma to the internal organs or bones can cause internal bleeding, which can sometimes be severe.
  - Dehydration can reduce the blood volume by reducing the water content of the blood. This would rarely result in shock (apart from the very severe cases) but may result in orthostatic hypotension and fainting.
- Disorders of circulation
  - Shock is the ineffective perfusion of tissues, and can be caused by a variety of conditions including blood loss, infection, poor cardiac output.
  - Atherosclerosis reduces the flow of blood through arteries, because atheroma lines arteries and narrows them. Atheroma tends to increase with age, and its progression can be compounded by many causes including smoking, high blood pressure, excess circulating lipids (hyperlipidemia), and diabetes mellitus.
  - Coagulation can form a thrombosis, which can obstruct vessels.
  - Problems with blood composition, the pumping action of the heart, or narrowing of blood vessels can have many consequences including hypoxia (lack of oxygen) of the tissues supplied. The term ischemia refers to tissue that is inadequately perfused with blood, and infarction refers to tissue death (necrosis), which can occur when the blood supply has been blocked (or is very inadequate)

HEMATOLOGICAL DISORDERS

- Anemia
  - Insufficient red cell mass (anemia) can be the result of bleeding, blood disorders like thalassemia, or nutritional deficiencies; and may require blood transfusion. Several countries have blood banks to fill the demand for transfusable blood. A person receiving a

blood transfusion must have a blood type compatible with that of the donor.

- Sickle-cell anemia

- Disorders of cell proliferation
  - Leukemia is a group of cancers of the blood-forming tissues and cells.
  - Non-cancerous overproduction of red cells (polycythemia vera) or platelets (essential thrombocytosis) may be premalignant.
  - Myelodysplastic syndromes involve ineffective production of one or more cell lines.
- Disorders of coagulation
  - Hemophilia is a genetic illness that causes dysfunction in one of the blood's clotting mechanisms. This can allow otherwise inconsequential wounds to be life-threatening, but more commonly results in hemarthrosis, or bleeding into joint spaces, which can be crippling.
  - Ineffective or insufficient platelets can also result in coagulopathy (bleeding disorders).
  - Hypercoagulable state (thrombophilia) results from defects in regulation of platelet or clotting factor function, and can cause thrombosis.
- Infectious disorders of blood
  - Blood is an important vector of infection. HIV, the virus that causes AIDS, is transmitted through contact with blood, semen or other body secretions of an infected person. Hepatitis B and C are transmitted primarily through blood contact. Owing to blood-borne infections, bloodstained objects are treated as a biohazard.
  - Bacterial infection of the blood is bacteremia or sepsis. Viral Infection is viremia. Malaria and trypanosomiasis are blood-borne parasitic infections.

CARBON MONOXIDE POISONING

Substances other than oxygen can bind to hemoglobin; in some cases this can cause irreversible damage to the body. Carbon monoxide, for example, is extremely dangerous when carried to the blood via the lungs by inhalation, because carbon monoxide irreversibly binds to hemoglobin to form carboxyhemoglobin, so that less hemoglobin is free to bind oxygen, and fewer oxygen molecules can be transported throughout the blood. This can cause suffocation insidiously. A fire burning in an enclosed room with poor ventilation presents a very dangerous hazard, since it can create a build-up of carbon monoxide in the air. Some carbon monoxide binds to hemoglobin when smoking tobacco.

MEDICAL TREATMENTS

BLOOD PRODUCTS

Blood for transfusion is obtained from human donors by blood donation and stored in a blood bank. There are many different blood types in humans, the ABO blood group system, and the Rhesus blood group system being the most important. Transfusion of blood of an incompatible blood group may cause severe, often fatal, complications, so crossmatching is done to ensure that a compatible blood product is transfused. Other blood products administered intravenously are platelets, blood plasma, cryoprecipitate, and specific coagulation factor concentrates.

INTRAVENOUS ADMINISTRATION

Many forms of medication (from antibiotics to chemotherapy) are administered intravenously, as they are not readily or adequately absorbed by the digestive tract.

After severe acute blood loss, liquid preparations, generically known as plasma expanders, can be given intravenously, either solutions of salts (NaCl, KCl, $CaCl_2$ etc.) at physiological concentrations, or colloidal solutions, such as dextrans, human serum albumin, or fresh frozen plasma. In these emergency situations, a plasma expander is a more effective life-saving procedure than a blood transfusion, because the metabolism of transfused red blood cells does not restart immediately after a transfusion.

BLOODLETTING

In modern evidence-based medicine, bloodletting is used in management of a few rare diseases, including hemochromatosis and polycythemia. However, bloodletting and leeching were common unvalidated interventions used until the 19th century, as many diseases were incorrectly thought to be due to an excess of blood, according to Hippocratic medicine.

2

Red Blood Cell

Red blood cells (RBCs), also called erythrocytes, are the most common type of blood cell and the vertebrate organism's principal means of delivering oxygen (O_2) to the body tissues—via blood flow through the circulatory system. RBCs take up oxygen in the lungs or gills and release it into tissues while squeezing through the body's capillaries.

The cytoplasm of erythrocytes is rich in hemoglobin, an iron-containing biomolecule that can bind oxygen and is responsible for the red color of the cells. The cell membrane is composed of proteins and lipids, and this structure provides properties essential for physiological cell function such as deformability and stability while traversing the circulatory system and specifically the capillary network.

In humans, mature red blood cells are flexible and oval biconcave disks. They lack a cell nucleus and most organelles, in order to accommodate maximum space for hemoglobin. Approximately 2.4 million new erythrocytes are produced per second in human adults. The cells develop in the bone marrow and circulate for about 100–120 days in the body before their components are recycled by macrophages. Each circulation takes about 20 seconds. Approximately a quarter of the cells in the human body are red blood cells.

Red blood cells are also known as RBCs, red cells, red blood corpuscles (an archaic term), haematids, erythroid cells or erythrocytes (from Greek erythros for "red" and kytos for "hollow vessel", with -cyte translated as "cell" in modern usage). Packed red blood cells (pRBC) are red blood cells that have been donated, processed, and stored in a blood bank for blood transfusion.

The first person to describe red blood cells was the young Dutch biologist Jan Swammerdam, who had used an early microscope in 1658 to study the blood of a frog. Unaware of this work, Anton van Leeuwenhoek provided another microscopic description in 1674, this time providing a more precise description of red blood cells, even approximating their size, "25,000 times smaller than a fine grain of sand".

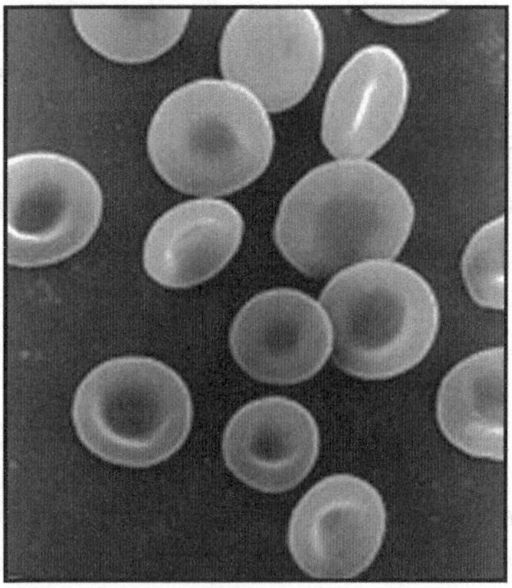

Fig. Scanning Electron Micrograph of Human Red Blood Cells (ca. 6–8 μm in Diameter.)

In 1901, Karl Landsteiner published his discovery of the three main blood groups—A, B, and C (which he later renamed to O). Landsteiner described the regular patterns in which reactions occurred when serum was mixed with red blood cells, thus identifying compatible and conflicting combinations between these blood groups. A year later Alfred von Decastello and Adriano Sturli, two colleagues of Landsteiner, identified a fourth blood group—AB. In 1959, by use of X-ray crystallography, Dr. Max Perutz was able to unravel the structure of hemoglobin, the red blood cell protein that carries oxygen. The oldest intact red blood cells ever discovered were found in Ötzi the Iceman, a natural mummy of a man who died around 3255 BCE. These cells were discovered in May 2012.

VERTEBRATE ERYTHROCYTES

Erythrocytes consist mainly of hemoglobin, a complex metalloprotein containing heme groups whose iron atoms temporarily bind to oxygen molecules (O_2) in the lungs or gills and release them throughout the body. Oxygen can easily diffuse through the red blood cell's cell membrane. Hemoglobin in the erythrocytes also carries some of the waste product carbon dioxide back from the tissues; most waste carbon dioxide, however, is transported back to the pulmonary capillaries of the lungs as bicarbonate (HCO_3^-) dissolved in the blood plasma. Myoglobin, a compound related to hemoglobin, acts to store oxygen in muscle cells.

The color of erythrocytes is due to the heme group of hemoglobin. The blood plasma alone is straw-colored, but the red blood cells change color depending on the state of the hemoglobin: when combined with oxygen the resulting oxyhemoglobin is scarlet, and when oxygen has been released the resulting deoxyhemoglobin is of a dark red burgundy color. However, blood can appear bluish when seen through the vessel wall and skin. Pulse oximetry takes advantage of the hemoglobin color change to directly measure the arterial blood oxygen saturation using colorimetric techniques. Hemoglobin also has a very high affinity for carbon monoxide, forming carboxyhemoglobin which is a very bright red in color. Flushed, confused patients with a saturation reading of 100% on pulse oximetry are sometimes found to be suffering from carbon monoxide poisoning.

The sequestration of oxygen-carrying proteins inside specialized cells (as opposed to oxygen carriers being dissolved in body fluid) was an important step in the evolution of vertebrates as it allows for less viscous blood, higher concentrations of oxygen, and better diffusion of oxygen from the blood to the tissues. The size of erythrocytes varies widely among vertebrate species; erythrocyte width is on average about 25% larger than capillary diameter, and it has been hypothesized that this improves the oxygen transfer from erythrocytes to tissues.

The only known vertebrates without erythrocytes are the crocodile icefishes (family Channichthyidae); they live in very oxygen-rich cold water and transport oxygen freely dissolved in their blood. While they do not use hemoglobin anymore, remnants of hemoglobin genes can be found in their genome.

NUCLEUS

Erythrocytes in mammals are anucleate when mature, meaning that they lack a cell nucleus. In comparison, the erythrocytes of other vertebrates have nuclei; the only known exceptions are salamanders of the Batrachoseps genus and fish of the Maurolicus genus with closely related species.

The elimination of the nucleus in vertebrate erythrocytes has been offered as an explanation for the subsequent accumulation of non-coding DNA in the genome. The argument runs as follows: Efficient gas transport requires erythrocytes to pass through very narrow capillaries, and this constrains their size. In the absence of nuclear elimination, the accumulation of repeat sequences is constrained by the volume occupied by the nucleus, which increases with genome size.

SECONDARY FUNCTIONS

When erythrocytes undergo shear stress in constricted vessels, they release ATP, which causes the vessel walls to relax and dilate so as to promote normal blood flow. When their hemoglobin molecules are deoxygenated, erythrocytes release S-nitrosothiols, which also act to dilate blood vessels, thus directing more blood to areas of the body depleted of oxygen. Erythrocytes can also synthesize nitric oxide enzymatically, using L-arginine as substrate, as do endothelial cells. Exposure of erythrocytes to physiological levels of shear stress activates nitric oxide synthase and export of nitric oxide, which may contribute to the regulation of vascular tonus. Erythrocytes can also produce hydrogen sulfide, a signalling gas that acts to relax vessel walls. It is believed that the cardioprotective effects of garlic are due to erythrocytes converting its sulfur compounds into hydrogen sulfide. Erythrocytes also play a part in the body's immune response: when lysed by pathogens such as bacteria, their hemoglobin releases free radicals, which break down the pathogen's cell wall and membrane, killing it.

MAMMALIAN ERYTHROCYTES

Mammalian erythrocytes are unique among the vertebrates as they are non-nucleated cells in their mature form. These cells have nuclei during early phases of erythropoiesis, but extrude them during development as they mature in order to provide more space for hemoglobin. In mammals, erythrocytes also lose all other cellular organelles such as their mitochondria, Golgi apparatus and endoplasmic reticulum.

As a result of not containing mitochondria, these cells use none of the oxygen they transport; instead they produce the energy carrier ATP by the glycolysis of glucose and lactic acid fermentation on the resulting pyruvate.

Because of the lack of nuclei and organelles, mature red blood cells do not contain DNA and cannot synthesize any RNA, and consequently cannot divide and have limited repair capabilities. This also ensures that no virus can evolve to target mammalian red blood cells.

Mammalian erythrocytes are typically shaped as biconcave disks: flattened and depressed in the center, with a dumbbell-shaped cross section, and a torus-shaped rim on the edge of the disk. This distinctive biconcave shape optimises the ow properties of blood in the large vessels, such as maximization of laminar flow and minimization of platelet scatter, which suppresses their atherogenic activity in those large vessels. However, there are some exceptions concerning shape in the artiodactyl order (even-toed ungulates including cattle, deer, and their relatives), which displays a wide variety of bizarre erythrocyte morphologies: small and highly ovaloid cells in llamas and camels (family Camelidae), tiny spherical cells in mouse deer (family Tragulidae), and cells which assume fusiform, lanceolate, crescentic, and irregularly polygonal and other angular forms in red deer and wapiti (family Cervidae). Members of this order have clearly evolved a mode of red blood cell development substantially different from the mammalian norm. Overall, mammalian erythrocytes are remarkably flexible and deformable so as to squeeze through tiny capillaries, as well as to maximize their apposing surface by assuming a cigar shape, where they efficiently release their oxygen load.

In large blood vessels, red blood cells sometimes occur as a stack, flat side next to flat side. This is known as rouleaux formation, and it occurs more often if the levels of certain serum proteins are elevated, as for instance during inflammation. The spleen acts as a reservoir of red blood cells, but this effect is somewhat limited in humans. In some other mammals such as dogs and horses, the spleen sequesters large numbers of red blood cells which are dumped into the blood during times of exertion stress, yielding a higher oxygen transport capacity.

HUMAN ERYTHROCYTES

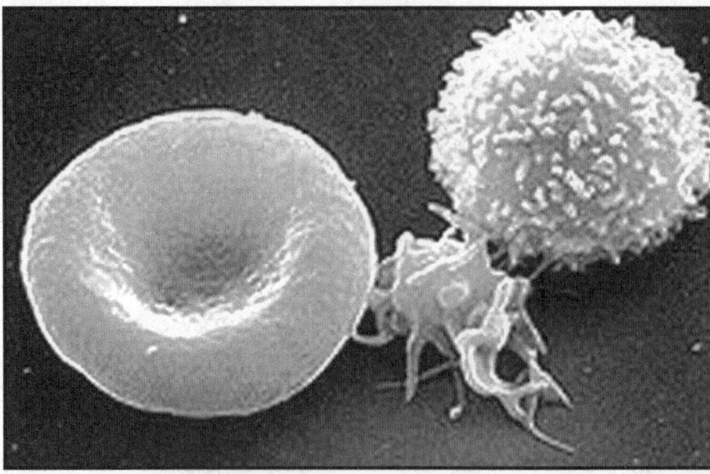

Fig. Scanning Electron Micrograph of Blood Cells. From Left to Right: Human Erythrocyte, Thrombocyte (Platelet), Leukocyte.

A typical human erythrocyte has a disk diameter of approximately 6.2–8.2 μm and a thickness at the thickest point of 2–2.5 μm and a minimum thickness in the centre of 0.8–1 μm, being much smaller than most other human cells. These cells have an average volume of about 90 fL with a surface of about 136 ìm^2, and can swell up to a sphere shape containing 150 fL, without membrane distension.

Adult humans have roughly 20–30 $\times$ 10^{12} (20–30 trillion) red blood cells at any given time, comprising approximately one quarter of the total human body cell number (women have about 4 to 5 million erythrocytes per microliter (cubic millimeter) of blood and men about 5 to 6 million; people living at high altitudes with low oxygen tension will have more). Red blood cells are thus much more common than the other blood particles: there are about 4,000–11,000 white blood cells and about 150,000–400,000 platelets in each microliter of human blood.

Human red blood cells take on average 20 seconds to complete one cycle of circulation.

As red blood cells contain no nucleus, protein biosynthesis is currently assumed to be absent in these cells, although a recent study indicates the presence of all the necessary biomachinery in the cells to do so.

The blood's red color is due to the spectral properties of the hemic iron ions in hemoglobin. Each human red blood cell contains approximately 270 million of these hemoglobin biomolecules, each carrying four heme groups; hemoglobin comprises about a third of the total cell volume. This protein is responsible for the transport of more than 98% of the oxygen (the remaining oxygen is carried dissolved in the blood plasma). The red blood cells of an average adult human male store collectively about 2.5 grams of iron, representing about 65% of the total iron contained in the body.

LIFE CYCLE

Human erythrocytes are produced through a process named erythropoiesis, developing from committed stem cells to mature erythrocytes in about 7 days. When matured, in a healthy individual these cells live in blood circulation for about 100 to 120 days (and 80 to 90 days in a full term infant). At the end of their lifespan, they become senescent, and are removed from circulation. In many chronic diseases, the lifespan of the erythrocytes is markedly reduced (e.g. patients requiring haemodialysis).

Erythropoiesis

Erythropoiesis is the development process by which new erythrocytes are produced; it lasts about 7 days. Through this process erythrocytes are continuously produced in the red bone marrow of large bones, at a rate of about 2 million per second in a healthy adult. (In the embryo, the liver is the main site of red blood cell production.) The production can be stimulated by the hormone erythropoietin (EPO), synthesised by the kidney. Just before and after leaving the bone marrow, the developing cells are known as reticulocytes; these comprise about 1% of circulating red blood cells.

Functional Lifetime

The functional lifetime of an erythrocyte is about 100–120 days, during which time the erythrocytes are continually moved by the blood flow push (in arteries), pull (in veins) and a combination of the two as they squeeze through microvessels such as capillaries.

Senescence

The aging erythrocyte undergoes changes in its plasma membrane, making it susceptible to selective recognition by macrophages and subsequent phagocytosis in the mononuclear phagocyte system (spleen, liver and lymph nodes), thus removing old and defective cells and continually purging the blood. This process is termed eryptosis, erythrocyte programmed cell death. This process normally occurs at the same rate of production by erythropoiesis, balancing the total circulating red blood cell count. Eryptosis is increased in a wide variety of diseases including sepsis, haemolytic uremic syndrome, malaria, sickle cell anemia, beta-thalassemia, glucose-6-phosphate dehydrogenase deficiency, phosphate depletion, iron deficiency and Wilson's disease. Eryptosis can be elicited by osmotic shock, oxidative stress, energy depletion as well as a wide variety of endogenous mediators and xenobiotics. Excessive eryptosis is observed in erythrocytes lacking the cGMP-dependent protein kinase type I or the AMP-activated protein kinase AMPK. Inhibitors of eryptosis include erythropoietin, nitric oxide, catecholamines and high concentrations of urea.

Much of the resulting breakdown products are recirculated in the body. The heme constituent of hemoglobin are broken down into Fe^{3+} and biliverdin. The biliverdin is reduced to bilirubin, which is released into the plasma and recirculated to the liver bound to albumin. The iron is released into the plasma to be recirculated by a carrier protein called transferrin. Almost all erythrocytes are removed in this manner from the circulation before they are old enough to hemolyze.

Hemolyzed hemoglobin is bound to a protein in plasma called haptoglobin, which is not excreted by the kidney.

MEMBRANE COMPOSITION

The membrane of the red blood cell plays many roles that aid in regulating their surface deformability, flexibility, adhesion to other cells and immune recognition. These functions are highly dependent on its composition, which defines its properties.

The red blood cell membrane is composed of 3 layers: the glycocalyx on the exterior, which is rich in carbohydrates; the lipid bilayer which contains many transmembrane proteins, besides its lipidic main constituents; and the membrane skeleton, a structural network of proteins located on the inner surface of the lipid bilayer. Half of the membrane mass in human and most mammalian erythrocytes are proteins. The other half are lipids, namely phospholipids and cholesterol.

Membrane Lipids

The erythrocyte cell membrane comprises a typical lipid bilayer, similar to what can be found in virtually all human cells. Simply put, this lipid bilayer is composed of cholesterol and phospholipids in equal proportions by weight. The lipid composition is important as it defines many physical properties such as membrane permeability and fluidity.

Additionally, the activity of many membrane proteins is regulated by interactions with lipids in the bilayer. Unlike cholesterol, which is evenly distributed between the inner and outer leaflets, the 5 major phospholipids are asymmetrically disposed, as shown below:

Outer Monolayer

Phosphatidylcholine

Phosphatidylcholines (PC) are a class of phospholipids that incorporate choline as a headgroup. They are a major component of biological membranes and can be easily obtained from a variety of readily available sources, such as egg yolk or soybeans, from which they are mechanically or chemically extracted using hexane. They are also a member of the lecithin group of yellow-brownish fatty substances occurring in animal and plant tissues. Dipalmitoyl phosphatidylcholine (aka: lecithin) is a major component of pulmonary surfactant and is often used in the L/S ratio to calculate fetal lung maturity. While phosphatidylcholines are found in all plant and animal cells, they are absent in the membranes of most bacteria, including Escherichia coli. Purified phosphatidylcholine is produced commercially.

The name "lecithin" was originally defined from the Greek lekithos (λεκιθοξ, egg yolk) by Theodore Nicolas Gobley, a French chemist and pharmacist of the mid-19th century, who applied it to the egg yolk phosphatidylcholine that he identified in 1847. Gobley eventually completely described his lecithin from chemical structural point of view, in 1874. Phosphatidylcholines are such a major component of lecithin that in some contexts the terms are sometimes used as synonyms. However, lecithin extracts consist of a mixture of phosphatidylcholine and other compounds. It is also used along with sodium taurocholate for simulating fed- and fasted-state biorelevant media in dissolution studies of highly lipophilic drugs.

Phosphatidylcholine is a major constituent of cell membranes and pulmonary surfactant, and is more commonly found in the exoplasmic or outer leaflet of a cell membrane. It is thought to be transported between membranes within the cell by phosphatidylcholine transfer protein (PCTP). Phosphatidylcholine also plays a role in membrane-mediated cell signaling and PCTP activation of other enzymes.

Structure and Physical Properties

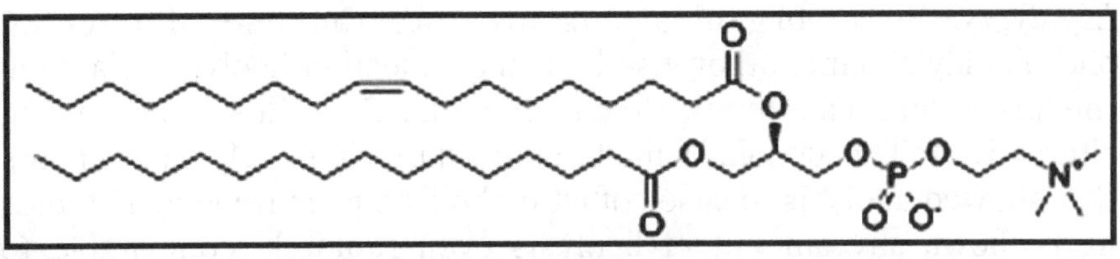

Fig. Palmitoyl-Oleyl-sn-Phosphatidylcholine, a Phosphatidylcholine Derivative.

This phospholipid is composed of a choline head group and glycerophosphoric acid, with a variety of fatty acids. Usually, one is a saturated fatty acid (in the given figure, this can be palmitic or hexadecanoic acid, H_3C-

$(CH_2)_{14}$-COOH; margaric acid identified by Gobley in egg yolk, or heptadecanoic acid H_3C-$(CH_2)_{15}$-COOH, also belong to that class); and the other is an unsaturated fatty acid (here oleic acid, or 9Z-octadecenoic acid, as in Gobley's original egg yolk lecithin).

However, there are also examples of disaturated species. Animal lung phosphatidylcholine, for example, contains a high proportion of dipalmitoylphosphatidylcholine.

Phospholipase D catalyzes the hydrolysis of phosphatidylcholine to form phosphatidic acid (PA), releasing the soluble choline headgroup into the cytosol. Phosphatidylcholine is a neutral lipid, but it carries an electric dipole moment of about 10 D. Vibrational dynamics of phosphatidylcholine and its hydration waters has been recently calculated from first principles.

Possible Health Benefits

- *Senescence:* Phosphatidylcholine is a vital substance found in every cell of the human body. Thus, some researchers have used mutant mouse models with severe oxidative damage as a model of "accelerated aging" to investigate the possible role of phosphatidylcholine supplementation as a way of slowing down aging-related processes and improving brain functioning and memory capacity in dementia. However, a systematic review of clinical trials in humans found that there is not enough evidence to support lecithin or phosphatidylcholine supplementation for patients with dementia. The study does admit that a moderate benefit cannot be ruled out until further large scale studies are performed.
- *Liver repair:* Recent studies have examined potential benefits of phosphatidylcholine for liver repair. Results are mixed in animal models, and no clinical evidence shows a health benefit in humans. One study shows the healing effect of phosphatidylcholine in mice with hepatitis A, hepatitis B, and hepatitis C. The administration of phosphatidylcholine for chronic, active hepatitis resulted in significant reduction of disease activity in mice.
- *Lipolysis:* Some organizations promote the use of injected phosphatidylcholine, otherwise known as injection lipolysis, claiming the procedure can break down fat cells, and thus serve as an alternative to liposuction. While the procedure cites early experiments that showed lipolysis in cases of fat emboli, no peer-reviewed studies have shown any amount of lipolysis even remotely comparable to liposuction.
- *Ulcerative colitis:* Phase IIa/b clinical trials performed at the Heidelberg University Hospital have shown that delayed release purified phosphatidylcholine is an anti-inflammatory agent, and a surface

hydrophobicity increasing compound with promising therapeutic potential in the treatment of inflammatory bowel disease.

Possible Health Risks

A report in 2011 has linked the microbial catabolites of phosphatidylcholine with increased atherosclerosis in mice through the production of choline, trimethylamine oxide, and betaine.

Sphingomyelin

Sphingomyelin is a type of sphingolipid found in animal cell membranes, especially in the membranous myelin sheath that surrounds some nerve cell axons. It usually consists of phosphocholine and ceramide, or a phosphoethanolamine head group; therefore, sphingomyelins can also be classified as sphingophospholipids. In humans, SPH represents ~85% of all sphingolipids, and typically make up 10-20 mol % of plasma membrane lipids.

Sphingomyelins contain phosphocholine or phosphoethanolamine as their polar head group and are therefore classified along with glycerophospholipids as phospholipids. Indeed, sphingomyelins resemble phosphatidylcholines in their general properties and three-dimensional structure, and in having no net charge on their head groups. Sphingomyelins are present in the plasma membranes of animal cells and are especially prominent in myelin, a membranous sheath that surrounds and insulates the axons of some neurons—thus the name "sphingomyelins.

Sphingomyelin was first isolated by German chemist Johann L.W. Thudicum in the 1880s. The structure of sphingomyelin was first reported in 1927 as N-acyl-sphingosine-1-phosphorylcholine. Sphingomyelin content in mammals ranges from 2 to 15% in most tissues, with higher concentrations found in nerve tissues, red blood cells, and the ocular lenses. Sphingomyelin has significant structural and functional roles in the cell. It is a plasma membrane component and participates in many signaling pathways. The metabolism of sphingomyelin creates many products that play significant roles in the cell.

Physical Characteristics

- *Composition:* Sphingomyelin consists of a phosphocholine head group, a sphingosine, and a fatty acid. It is one of the few membrane phospholipids not synthesized from glycerol. The sphingosine and fatty acid can collectively be categorized as a ceramide. This composition allows sphingomyelin to play significant roles in signaling pathways: the degradation and synthesis of sphingomyelin produce important second messengers for signal transduction. Sphingomyelin obtained from natural sources, such as eggs or bovine brain, contains fatty acids of various chain length. Sphingomyelin with set chain length, such as palmitoylsphingomyelin with a saturated 16 acyl chain, is available commercially.

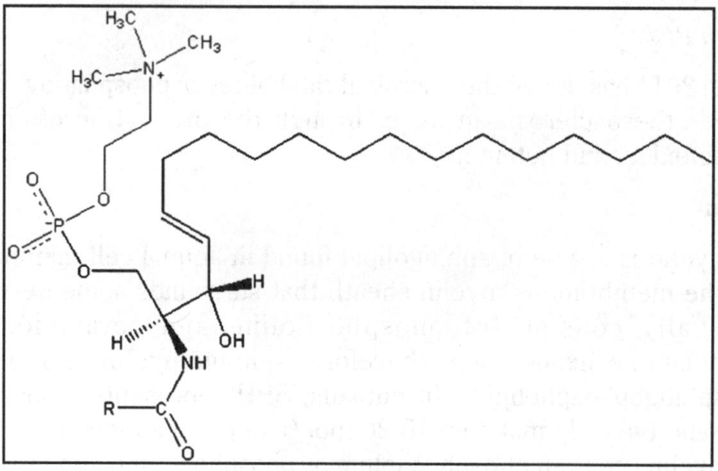

Fig. Sphingomyelin.

- *Properties:* Ideally, sphingomyelin molecules are shaped like a cylinder, however many molecules of sphingomyelin have a significant chain mismatch (the lengths of the two hydrophobic chains are significantly different). The hydrophobic chains of sphingomyelin tend to be much more saturated than other phospholipids. The main transition phase temperature of sphingomyelins is also higher compared to the phase transition temperature of similar phospholipids, near 37 C. This can introduce lateral heterogeneity in the membrane, generating domains in the membrane bilayer. Sphingomyelin undergoes significant interactions with cholesterol. Cholesterol has the ability to eliminate the liquid to solid phase transition in phospholipids. Due to sphingomyelin transition temperature being within physiological temperature ranges, cholesterol can play a significant role in the phase of sphingomyelin. Sphingomyelin are also more prone to intermolecular hydrogen bonding than other phospholipids.
- *Location:* Sphingomyelin is synthesized at the endoplasmic reticulum (ER), where it can be found in low amounts, and at the trans Golgi. It is enriched at the plasma membrane with a greater concentration on the outer than the inner leaflet. The Golgi complex represents an intermediate between the ER and plasma membrane, with slightly higher concentrations towards the trans side.

Metabolism

The synthesis of sphingomyelin involves the enzymatic transfer of a phosphocholine from phosphatidylcholine to a ceramide. The first committed step of sphingomyelin synthesis involves the condensation of L-serine and palmitoyl-CoA. This reaction is catalyzed by serine palmitoyltransferase. The product of this reaction is reduced, yielding dihydrosphingosine. The dihydrosphingosine undergoes N-acylation followed by desaturation to yield a ceramide. Each one of these reactions occurs at the cytosolic surface of the endoplasmic reticulum. The ceramide is transported to the Golgi apparatus where it can be converted to sphingomyelin. Sphingomyelin synthase is responsible for the production of sphingomyelin from ceramide. Diacylglycerol is produced as a byproduct when the phosphocholine is transferred.

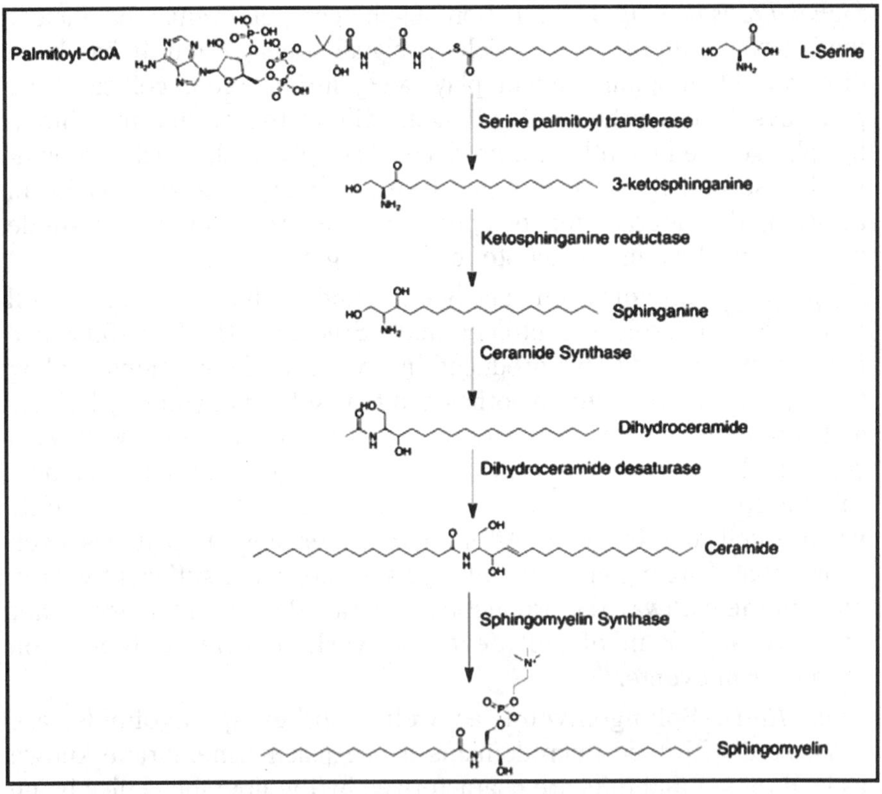

Fig. Sphingomyelin de Novo Synthesis Pathway.

Sphingomyelin breakdown is responsible for initiating many universal signaling pathways. It is hydrolyzed by sphingomyelinases (sphingomyelin specific type-C phosopholipases). The phosphocholine head group is released into the aqueous environment while the ceramide diffuses through the membrane.

Function

- *Membranes:* The membranous myelin sheath that surrounds and electrically insulates many nerve cell axons is particularly rich in sphingomyelin, suggesting its role as an insulator of nerve fibers. The plasma membrane of other cells is also abundant in sphingomyelin, though it is largely to be found in the exoplasmic leaflet of the cell membrane. There is, however, some evidence that there may also be a sphingomyelin pool in the inner leaflet of the membrane. Moreover, neutral sphingomyelinase-2 – an enzyme that breaks down sphingomyelin into ceramide - has been found to localise exclusively to the inner leaflet, further suggesting that there may be sphingomyelin present there.

- *Signal Transduction:* The function of sphingomyelin remained unclear until it was found to have a role in signal transduction. It has been discovered that sphingomyelin plays a significant role in cell signaling pathways. The synthesis of sphingomyelin at the plasma membrane by sphingomyelin synthase 2 produces diacylglycerol, which is a lipid-soluble second messenger that can pass along a signal cascade. In addition, the degradation of sphingomyelin can produce ceramide which is involved in the apoptotic signalling pathway.
- *Apoptosis:* Sphingomyelin has been found to have a role in cell apoptosis by hydrolyzing into ceramide. Studies in the late 1990s had found that ceramide was produced in a variety of conditions leading to apoptosis. It was then hypothesized that sphingomyelin hydrolysis and ceramide signaling were essential in the decision of whether a cell dies. In the early 2000s new studies emerged that defined a new role for sphingomyelin hydrolysis in apoptosis, determining not only when a cell dies but how. After more experimentation it has been shown that if sphingomyelin hydrolysis happens at a sufficiently early point in the pathway the production of ceramide may influence either the rate and form of cell death or work to release blocks on downstream events.
- *Lipid Rafts:* Sphingomyelin, as well as other sphingolipids, are associated with lipid microdomains in the plasma membrane known as lipid rafts. Lipid rafts are characterized by the lipid molecules being in the lipid ordered phase, offering more structure and rigidity compared to the rest of the plasma membrane. In the rafts, the acyl chains have low chain motion but the molecules have high lateral mobility. This order is in part due to the higher transition temperature of sphingolipids as well as the interactions of these lipids with cholesterol. Cholesterol is a relatively small, nonpolar molecule that can fill the space between the sphingolipids that is a result of the large acyl chains. Lipid rafts are thought to be involved in many cell processes, such as membrane sorting and trafficking, signal transduction, and cell polarization. Due to the specific types of lipids

in these microdomains, lipid rafts can accumulate certain types of proteins associated with them, thereby increasing the special functions they possess. Lipid rafts have been speculated to be involved in the cascade of cell apoptosis.

Abnormalities and Associated Diseases

Sphingomyelin can accumulate in a rare hereditary disease called Niemann-Pick Disease, types A and B. It is a genetically-inherited disease caused by a deficiency in the enzyme sphingomyelinase, which causes the accumulation of sphingomyelin in spleen, liver, lungs, bone marrow, and brain, causing irreversible neurological damage.

Of the two types involving sphingomyelinase, type A occurs in infants. It is characterized by jaundice, an enlarged liver, and profound brain damage. Children with this type rarely live beyond 18 months. Type B involves an enlarged liver and spleen, which usually occurs in the pre-teen years. The brain is not affected. Most patients present with $<1\%$ normal levels of the enzyme in comparison to normal levels.

As a result of the autoimmune disease multiple sclerosis (MS), the myelin sheath of neuronal cells in the brain and spinal cord is degraded, resulting in loss of signal transduction capability. MS patients exhibit upregulation of certain cytokines in the cerebrospinal fluid, particularly tumor necrosis factor alpha. This activates sphingomyelinase, an enzyme that catalyzes the hydrolysis of sphingomyelin to ceramide; sphingomyelinase activity has been observed in conjunction with cellular apoptosis.

An excess of sphingomyelin in the red blood cell membrane (as in abetalipoproteinemia) causes excess lipid accumulation in the outer leaflet of the red blood cell plasma membrane. This results in abnormally shaped red cells called acanthocytes.

Inner Monolayer

Phosphatidylethanolamine

Phosphatidylethanolamines (sometimes abbreviated PE) are a class of phospholipids found in biological membranes. They are synthesized by the addition of CDP-ethanolamine to diglycerides, releasing CMP. S-Adenosyl methionine can subsequently methylate the amine of phosphatidylethanolamines to yield phosphatidylcholines. It can mainly be found in the inner (cytoplasmic) leaflet of the lipid bilayer.

Function

- *In cells:* Phosphatidylethanolamines are found in all living cells, composing 25% of all phospholipids. In human physiology, they are found particularly in nervous tissue such as the white matter of brain, nerves, neural tissue, and in spinal cord, where they make up 45% of all phospholipids. PEs play a role in membrane fusion and in disassembly of the contractile ring during cytokinesis in cell division. Additionally, it is thought that PE regulates membrane curvature. PE acts as an important precursor, substrate, or donor in several biological pathways. As a polar head group, PE creates a more viscous lipid membrane compared to phosphatidylcholine (PC). For example, the melting temperature of di-oleoyl-PE is -16 °C while the melting temperature of di-oleoyl-PC is -20 °C. If the lipids had two palmitoyl chains, PE would melt at 63°C while PC would melt already at 41 °C. Lower melting temperatures correspond, in a simplistic view, to more fluid membranes.

- *In humans:* In humans, metabolism of PE is thought to be important in the heart. When blood flow to the heart is restricted, the asymmetrical distribution of PE between membrane leaflets is disrupted, and as a result the membrane is disrupted. Additionally, PE plays a role in the secretion of lipoproteins in the liver. This is because vesicles for secretion of VLDLs coming off of the Golgi have a significantly higher PE concentration when compared to other vesicles containing VLDLs. PE has also shown to be able to propagate infectious prions without the assistance of any proteins or nucleic acids, which is a unique characteristic of it. PE is also thought to play a role in blood clotting, as it works with phosphatidylserine to increase the rate of thrombin formation by promoting binding to Factor V and Factor X, two proteins which catalyze the formation of thrombin from prothrombin.
- *In bacteria:* Where phosphatidylcholine is the principal phospholipid in animals, PE is the principal one in bacteria. One of the primary roles for PE in bacterial membranes is to spread out the negative charge caused by anionic membrane phospholipids. In the bacterium E. coli, PE play a role in supporting lactose permease's active transport of lactose into the cell, and may play a role in other transport systems as well. PE plays a role in the assembly of lactose permease and other membrane proteins. It acts as a 'chaperone' to help the membrane proteins correctly fold their tertiary structures so that they can function properly. When PE is not present, the transport proteins have incorrect tertiary structures and do not function correctly. PE also enables bacterial multidrug transporters to function properly. PE allows the formation of intermediates that are needed for the transporters to properly open and close.

Structure

As a lecithin, PE consists of a combination of glycerol esterified with two fatty acids and phosphoric acid. Whereas the phosphate group is combined with choline in phosphatidylcholine, it is combined with the ethanolamine in PE. The two fatty acids may be the same, or different, and are usually in the 1,2 positions (though they can be in the 1,3 positions).

Synthesis

The phosphatidylserine decarboxylation pathway and the CDP-ethanolamine pathways are used to synthesize PE. Phosphatidylserine decarboxylase (PSD) is the enzyme that is used to decarboxylate phosphatidylserine in the first pathway.

The phosphatidylserine decarboxylation pathway is the main source of synthesis for PE in the membranes of the mitochondria. PE produced in the mitochondrial membrane is also transported throughout the cell to other membranes for use. In a process that mirrors phosphatidylcholine synthesis, PE is also made via the CDP-ethanolamine pathway, using ethanolamine as the substrate. Through several steps taking place is both the cytosol and endoplasmic reticulum, the synthesis pathway yields the end product of PE. PE is also found abundantly in soya or egg lecithin and is produced commercially using chromatographic separation.

Synthesis of PE through the phosphatidylserine decarboxylation pathway occurs rapidly in the inner mitochondrial membrane. However, phosphatidylserine is made in the endoplasmic reticulum. Because of this, the transport of phosphatidylserine from the endoplasmic reticulum to the mitrochondrial membrane and then to the inner mitochondrial membrane limits the rate of synthesis via this pathway. The mechanism for this transport is currently unknown, but may play a role in regulation of the rate of synthesis in this pathway.

Presence in Food, Health Issues

Phosphatidylethanolamines in food break down to form PE-linked Amadori products as a part of the Maillard reaction. These products accelerate membrane lipid peroxidation, causing oxidative stress to cells that come in contact with them. Oxidative stress is known to cause food deterioration and several diseases. Significant levels of Amadori-PE products have been found in a wide variety of foods such as chocolate, soybean milk, infant formula, and other processed foods.

The levels of Amadori-PE products are higher in foods with high lipid and sugar concentrations that have high temperatures in processing. Additional studies have found that Amadori-PE may play a role in vascular disease, act as the mechanism by which diabetes can increase the incidence of cancer, and potentially play a role in other diseases as well. Amadori-PE has a higher plasma concentration in diabetes patients than healthy people, indicating it may play a role in the development of the disease or be a product of the disease.

Inositol Phosphate

Inositol phosphates are a group of mono- to polyphosphorylated inositols. They play crucial roles in diverse cellular functions, such as cell growth, apoptosis, cell migration, endocytosis, and cell differentiation.

The group comprises:

- Inositol monophosphate (IP)
- Inositol bisphosphate (IP_2)
- Inositol trisphosphate (IP_3)
- Inositol pentakisphosphate (IP_5)
- Inositol hexaphosphate (IP_6) also known as phytic acid, or phytate (as a salt).

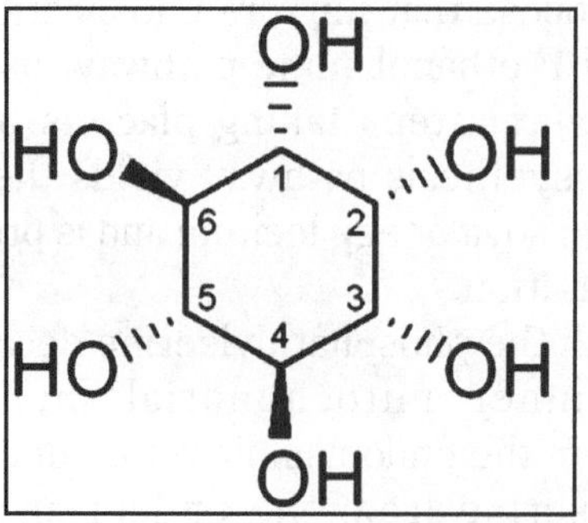

Fig. Inositol.

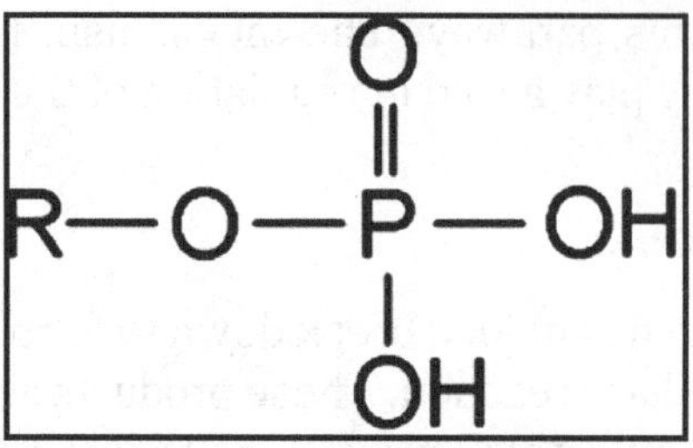

Fig. Phosphate Group.

Functions

- *Inositol triphosphate:* Inositol trisphosphates act on the inositol triphosphate receptor to release calcium into the cytoplasm. Further reading: Function of calcium in humans
- *Other:* Inositol tetra-, penta-, and hexa-phosphates have been implicated in gene expression and Steger (both in Science Magazine).

Phosphatidylserine

Phosphatidylserine (abbreviated Ptd-L-Ser or PS) is an important phospholipid membrane component (i.e. component of the cell membrane) which plays a key role in cell cycle signaling, specifically in relationship to apoptosis.

Cell Signaling

Phosphatidylserine(s) are actively held facing the cytosolic (inner) side of the cell membrane by the enzyme flippase. This is in contrast to normal behavior of phospholipids in the cell membrane which can freely flip their heads between the two faces of the membrane they comprise. However, when a cell undergoes apoptosis phosphatidylserine is no longer restricted to the cytosolic domain by flippase. When the phosphatidylserines naturally flip to the extracellular (outer) surface of the cell, they act as a signal for macrophages to engulf the cells.

Supplementation and Health Benefits

- *Memory and cognition:* In May, 2003 the Food and Drug Administration gave "qualified health claim" status to phosphatidylserine thus allowing labels to state "consumption of phosphatidylserine may reduce the risk of dementia and cognitive dysfunction in the elderly" along with the disclaimer "very limited and preliminary scientific research suggests that phosphatidylserine may reduce the risk of cognitive dysfunction in the elderly. FDA concludes that there is little scientific evidence supporting this claim." The FDA declared that "based on its evaluation of the totality of the publicly available scientific evidence, the agency concludes that there is not significant scientific agreement among qualified experts that a relationship exists between phosphatidylserine and reduced risk of dementia or cognitive dysfunction". The FDA also noted "Of the 10 intervention studies that formed the basis of FDA's evaluation, all were seriously flawed or limited in their reliability in one or more ways", concluding that "most of the evidence does not support a relationship between phosphatidylserine and reduced risk of dementia or cognitive dysfunction, and that the evidence that does support such a relationship is very limited and preliminary". Early studies of phosphatidylserine on memory and cognition used a supplement which isolated the molecule from the bovine brain. Currently, most commercially available products are made from cabbage or soybeans due to the risk of mad cow disease in bovine brain tissue. These plant-based products have a similar, but not identical chemical structure to the bovine derived supplements; for example, the FDA notes "the phosphatidylserine molecule from soy lecithin contains mainly polyunsaturated acids, while the phosphatidylserine molecule from bovine brain cortex contains mainly saturated and monounsaturated fatty acids and long-chain polyunsaturated fatty acids". A preliminary study in rats in 1999 indicated that the soy derived phosphatidylserine supplement was as effective as the bovine derived supplement in one of three behavioral tests. However, clinical trials in humans found that "a daily supplement of S-PS [soybean-derived phosphatidylserine] does not affect memory or other cognitive functions in older individuals with memory complaints."

- *Sports nutrition:* In athletes, phosphatidylserine has been shown to improve performance, endocrine response to exercise stress, and decrease muscle damage in athletes involved in cycling, weight training, golf and endurance running. PS has been reported to be an effective supplement for combating exercise-induced stress by blunting the exercise-induced increase in cortisol levels in a dose dependent manner. PS supplementation promotes a desirable hormonal balance for athletes and might attenuate the physiological deterioration that accompanies overtraining and/or overstretching. In recent studies, PS has been shown to enhance mood in a cohort of young people during mental stress and to improve accuracy during tee-off by increasing the stress resistance of golfers.

Safety

Traditionally, PS supplements were derived from bovine cortex (BC-PS). However, due to the risk of potential transfer of infectious diseases, soy-derived PS (S-PS) supplements have been used as an alternative. Soy-derived PS is designated Generally Recognized As Safe by the FDA. A 2002 safety report determined supplementation in elder people at a dosage of 200 mg three times daily to be safe.

Target for Tumor Treatment

Bavituximab is a monoclonal antibody directed against phosphatidylserine (PS) expressed on tumor endothelium. Various pathophysiologic processes cause the exposure of PS on the external membrane leaflet. Bavituximab, once bound, starts up host effector activities, such as antibody dependent cellular cytotoxicity, causing vessel destruction and enhancing antitumor immunity. Specifically, bavituximab binds PS when recognizing a complex formed by PS and two cross-linked molecules of the PS-binding protein beta 2 glycoprotein-1 (beta 2GP1). Bavituximab is not currently an approved drug, but it has shown promising survival results in previously treated non-small cell lung cancer and other indications.

Dietary Sources

The average daily phosphatidylserine (PS) intake from diet in Western countries is estimated to be 130 mg. PS may be found in meat and fish. Only small amounts of PS can be found in dairy products or in vegetables, with the exception of white beans and soy lecithin.

Table. PS Content in Different Foods.

Food	PS Content in mg/100 g
Soy lecithin	5900
Bovine brain	713
Atlantic mackere	1480
Chicken heart	414
Atlantic herring	360
Eel	335
Offal (average value)	305
Pig's spleen	239
Pig's kidney	218
Tuna	194
Chicken leg, with skin, without bone	134
Chicken liver	123
White beans	107
Soft-shell clam	87
Chicken breast, with skin	85
Mullet	76
Veal	72
Beef	69
Pork	57
Pig's liver	50
Turkey leg, without skin or bone	50
Turkey breast without skin	45
Crayfish	40
Cuttlefish	31
Atlantic cod	28
Anchovy	25
Whole grain barley	20
European hake	17
European pilchard (sardine)	16
Trout	14
Rice (unpolished)	3
Carrot	2
Ewe's Milk	2
Cow's Milk (whole, 3.5% fat)	1
Potato	1

Metabolism

Phosphatidylserine is biosynthesized in bacteria by condensing the amino acid serine with CDP (cytidine diphosphate)-activated phosphatidic acid. In mammals, phosphatidylserine is produced by base-exchange reactions with phosphatidylcholine and phosphatidylethanolamine. Conversely, phosphatidylserine can also give rise to phosphatidylethanolamine and phosphatidylcholine, although in animals the pathway to generate phosphatidylcholine from phosphatidylserine only operates in the liver.

This asymmetric phospholipid distribution among the bilayer is the result of the function of several energy-dependent and energy-independent phospholipid transport proteins. Proteins called "Flippases" move phospholipids from the outer to the inner monolayer, while others called "floppases" do the opposite operation, against a concentration gradient in an energy dependent manner. Additionally, there are also "scramblase" proteins that move phospholipids in both directions at the same time, down their concentration gradients in an energy independent manner. There is still considerable debate ongoing regarding the identity of these membrane maintenance proteins in the red cell membrane.

The maintenance of an asymmetric phospholipid distribution in the bilayer (such as an exclusive localization of PS and PIs in the inner monolayer) is critical for the cell integrity and function due to several reasons:

- Macrophages recognize and phagocytose red cells that expose PS at their outer surface. Thus the confinement of PS in the inner monolayer is essential if the cell is to survive its frequent encounters with macrophages of the reticuloendothelial system, especially in the spleen.
- Premature destruction of thallassemic and sickle red cells has been linked to disruptions of lipid asymmetry leading to exposure of PS on the outer monolayer.
- An exposure of PS can potentiate adhesion of red cells to vascular endothelial cells, effectively preventing normal transit through the microvasculature. Thus it is important that PS is maintained only in the inner leaflet of the bilayer to ensure normal blood flow in microcirculation.
- Both PS and phosphatidylinositol-4,5-bisphosphate (PIP2) can regulate membrane mechanical function, due to their interactions with skeletal proteins such as spectrin and protein 4.1R. Recent studies have shown that binding of spectrin to PS promotes membrane mechanical stability. PIP2 enhances the binding of protein band 4.1R to glycophorin C but decreases its interaction with protein band 3, and thereby may modulate the linkage of the bilayer to the membrane skeleton.

The presence of specialized structures named "lipid rafts" in the erythrocyte membrane have been described by recent studies. These are structures enriched in cholesterol and sphingolipids associated with specific membrane proteins, namely flotillins, stomatins (band 7), G-proteins, and â-adrenergic receptors. Lipid rafts that have been implicated in cell signaling events in nonerythroid cells have been shown in erythroid cells to mediate â2-adregenic receptor signaling and increase cAMP levels, and thus regulating entry of malarial parasites into normal red cells.

Membrane Proteins

The proteins of the membrane skeleton are responsible for the deformability, flexibility and durability of the red blood cell, enabling it to squeeze through capillaries less than half the diameter of the erythrocyte (7—8μm) and recovering the discoid shape as soon as these cells stop receiving compressive forces, in a similar fashion to an object made of rubber.

There are currently more than 50 known membrane proteins, which can exist in a few hundred up to a million copies per erythrocyte. Approximately 25 of these membrane proteins carry the various blood group antigens, such as the A, B and Rh antigens, among many others. These membrane proteins can perform a wide diversity of functions, such as transporting ions and molecules across the red cell membrane, adhesion and interaction with other cells such as endothelial cells, as signaling receptors, as well as other currently unknown functions. The blood types of humans are due to variations in surface glycoproteins of erythrocytes. Disorders of the proteins in these membranes are associated with many disorders, such as hereditary spherocytosis, hereditary elliptocytosis, hereditary stomatocytosis, and paroxysmal nocturnal hemoglobinuria. The red blood cell membrane proteins organized according to their function:

TRANSPORT

Band 3

Band 3 anion transport protein also known as anion exchanger 1 (AE1) or band 3 or solute carrier family 4 member 1 (SLC4A1) is a protein that in humans is encoded by the SLC4A1 gene. Band 3 anion transport protein is a phylogenetically preserved transport protein responsible for mediating the exchange of chloride (Cl^-) for bicarbonate (HCO_3^-) across a plasma membrane. Functionally similar members of the AE clade are AE2 and AE3.

Function

This is present in the principal acid secreting cell of the kidney, which generates hydrogen ions and bicarbonate ions from carbon dioxide and water - a reaction catalysed by Carbonic anhydrase. The hydrogen ions are pumped into the collecting duct tubule by vacuolar H^+ ATPase, the apical proton pump, which thus excretes acid into the urine. kAE1 exchanges bicarbonate for chloride on the basolateral surface, essentially returning bicarbonate to the blood.

Here it performs two functions:

- Electroneutral chloride and bicarbonate exchange across the plasma membrane on a one-for-one basis.This is crucial for CO_2 uptake by the red blood cell and conversion (by hydration catalysed by carbonic anhydrase) into a proton and a bicarbonate ion. The bicarbonate is then extruded (in exchange for a chloride) from the cell by the band 3 molecule.
- Physical linkage of the plasma membrane to the underlying membrane skeleton (via binding with ankyrin and protein 4.2). This appears to be to prevent membrane surface loss, rather than being to do with membrane skeleton assembly.

Species Distribution

It is ubiquitous throughout the vertebrates. In humans it is present in two specific sites:

- The erythrocyte (red blood cell) cell membrane and
- The basolateral surface of the alpha-intercalated cell (the acid secreting cell type) in the collecting duct of the kidney.

Tissue Distribution

The erythrocyte and kidney forms are different isoforms of the same protein. AE1 is an important structural component of the erythrocyte cell membrane, making up to 25% of the cell membrane surface.

Each red cell contains approximately one million copies of AE1. A different isoform of AE1, known as kAE1 (which is 65 amino acids shorter than erythroid AE1) is found in the basolateral surface of the alpha-intercalated cell in the cortical collecting duct of the kidney.

Clinical Significance

Mutations of kidney AE1 cause distal (type1) renal tubular acidosis, which is an inability to acidify the urine, even if the blood is too acidic. These mutations are disease causing as they cause mistargetting of the mutant band 3 proteins so that they are retained within the cell or occasionally addressed to the wrong (i.e. apical) surface.

Mutations of erythroid AE1 affecting the extracellular domains of the molecule may cause alterations in the individual's blood group, as band 3 determines the Diego blood group.

More importantly erythroid AE1 mutations cause 15–25% of cases of Hereditary spherocytosis (a disorder associated with progressive red cell membrane loss), and also cause the hereditary conditions of Hereditary stomatocytosis and Southeast Asian Ovalocytosis

Interactions

Band 3 has been shown to interact with CA2 and CA4.

Discovery

AE1 was discovered following SDS-PAGE gel electrophoresis of erythrocyte cell membrane. The large 'third' band on the electrophoresis gel represented AE1, which was thus initially termed 'Band 3'. The chloride-bicarbonate exchanger in the red cell membrane is not a pump, which would use metabolic energy. Nor is it strictly an enzyme. It is protein counter-transporter, known as band III.

Aquaporin 1

Aquaporin 1 is a protein that in humans is encoded by the AQP1 gene. AQP1 is a widely expressed water channel, whose physiological function has been most thoroughly characterized in the kidney. It is found in the basolateral and apical plasma membranes of the proximal tubules, the descending limb of the loop of Henle, and in the descending portion of the vasa recta. Additionally, it is found in red blood cells, vascular endothelium, the gastrointestinal tract, sweat glands, and lungs. It is not regulated by vasopressin (ADH).

Function

Aquaporins are a family of small integral membrane proteins related to the major intrinsic protein (MIP or AQP0). This gene encodes an aquaporin which functions as a molecular water channel protein. It is a homotetramer with 6 bilayer spanning domains and N-glycosylation sites. The protein physically resembles channel proteins and is abundant in erythrocytes and renal tubes. The gene encoding this aquaporin is a possible candidate for disorders involving imbalance in ocular fluid movement.

Glut1

Glucose transporter 1 (or GLUT1), also known as solute carrier family 2, facilitated glucose transporter member 1 (SLC2A1), is a uniporter protein that in humans is encoded by the SLC2A1 gene. GLUT1 facilitates the transport of glucose across the plasma membranes of mammalian cells.

Discovery

GLUT1 was the first glucose transporter to be characterized. GLUT 1 is highly conserved. GLUT 1 of humans and mouse has 98% homology. GLUT 1 has 40% homology with other GLUTs.

Function

Energy-yielding metabolism in erythrocytes depends on a constant supply of glucose from the blood plasma, where the glucose concentration is maintained at about 5mM. Glucose enters the erythrocyte by facilitated diffusion via a specific glucose transporter, at a rate about 50,000 times greater than uncatalyzed transmembrane diffusion. The glucose transporter of erythrocytes (called GLUT1 to distinguish it from related glucose transporters in other tissues) is a type III integral protein with 12 hydrophobic segments, each of which is believed to form a membrane-spanning helix. The detailed structure of GLUT1 is not known yet, but one plausible model suggests that the side-by-side assembly of several helices produces a transmembrane channel lined with hydrophilic residues that can hydrogen-bond with glucose as it moves through the channel.

GLUT1 is responsible for the low level of basal glucose uptake required to sustain respiration in all cells. Expression levels of GLUT1 in cell membranes are increased by reduced glucose levels and decreased by increased glucose levels.

GLUT1 is also a major receptor for uptake of Vitamin C as well as glucose, especially in non vitamin C producing mammals as part of an adaptation to compensate by participating in a Vitamin C recycling process. In mammals that do produce Vitamin C, GLUT4 is often expressed instead of GLUT1.

Tissue Distribution

It is widely distributed in fetal tissues. In the adult it is expressed at highest levels in erythrocytes and also in the endothelial cells of barrier tissues such as the blood–brain barrier.

Structure

GLUT1 behaves as a Michaelis-Menten enzyme and contains 12 membrane-spanning alpha helices, each containing 20 amino acid residues. A helical wheel analysis shows that the membrane spanning alpha helices are amphipathic, with one side being polar and the other side hydrophobic. Six of these membrane spanning helices are believed to bind together in the membrane to create a polar channel in the center through which glucose can traverse, with the hydrophobic regions on the outside of the channel adjacent to the fatty acid tails of the membrane.

Clinical Significance

Mutations in the GLUT1 gene are responsible for GLUT1 deficiency or De Vivo disease, which is a rare autosomal dominant disorder. This disease is characterized by a low cerebrospinal fluid glucose concentration (hypoglycorrhachia), a type of neuroglycopenia, which results from impaired glucose transport across the blood–brain barrier. GLUT1 is also a receptor used by the HTLV virus to gain entry into target cells. Glut1 has also been demonstrated as a powerful histochemical marker for haemangioma of infancy

Interactions

GLUT1 has been shown to interact with GIPC1. GLUT1 has two significant types in brain 45k and 55k. GLUT1 45k is present on astroglia of neurons and GLUT1 55k is present on capillaries in brain and is responsible for glucose transport across blood brain barrier and its deficiency causes low level of glucose in CSF(less than 60 mg/dl) which may manifest as convulsion in deficient individuals. Recently it has been described a GLUT1 inhibitor, DERL3, that is often methylated in colorectal cancer. In this cancer, DERL3 methylations seems to mediate the Warburg Effect.

Inhibitors

Fasentin is a small molecule inhibitor of the intracellular domain of GLUT1 preventing glucose uptake.

Kidd Antigen System

The Kidd antigen system (also known as Jk antigen) is present on the membranes of red blood cells and the kidney and helps determine a person's blood type. The Jk antigen is found on a protein responsible for urea transport in the red blood cells and the kidney. The gene encoding this protein is found on chromosome 18.

Three Jk alleles are Jk(a), Jk(b)and Jk3. Jk (a) was discovered by Allen et al. in 1951 and is named after a patient (Mrs Kidd delivered a baby with a haemolytic disease of the newborn associated with an antibody directed against a new antigen Jka). Whereas Jk (b) was discovered by Plant et al. in 1953, individuals who lack the Jk antigen (Jk null) are unable to maximally concentrate their urine.

The Jk antigen is important in transfusion medicine. People with two Jk(a) antigens, for instance, may form antibodies against donated blood containing two Jk(b) antigens (and thus no Jk(a) antigens). This can lead to hemolytic anemia, in which the body destroys the transfused blood, leading to low red blood cell counts.

Another disease associated with the Jk antigen is hemolytic disease of the newborn (HDN), in which a pregnant woman's body creates antibodies against the blood of her fetus, leading to destruction of the fetal blood cells. HDN associated with Jk antibodies is typically mild, though fatal cases have been reported.

Genetics and Biochemistry

Kidd comprises three antigens on a glycoprotein with 10 transmembrane spanning domains, cytoplasmic N- and C-termini and one extracellular N-glycosylation site. The Kidd gene has 11 exons with exons 4-11 encoding the mature protein. The Kidd gene (SLC14A1) is on chromosome 18q12.3.

Kidd antigens

Jka (JK1) and Jkb (JK2)

Jka and Jkb are the products of alleles with Asp280 and Asn280 in the fourth external loop of the Kidd glycoprotein. Jka and Jkb have similar prevalences in White and Asian populations but Jka is more common in Blacks than Jkb. Kidd antigens are resistant to the proteolytic enzymes papain and ficin.

Jk(a-b-) and Jk3

- Jk(a-b-) represents the null phenotype and usually results from homozygosity for a silent gene at the JK locus. The null phenotype is rare in most populations but does have increased prevalence in Polynesians (one in 400) and Niueans (1.4%).
- In Polynesians the null allele contains a splice site mutation in intron 5 causing a loss of exon 6 from the mRNA product.
- In Finns (null phenotype less rare than in other European populations), the null phenotype results from a mutation encoding a Ser291Pro substitution.
- A rare null phenotype in Japanese individuals results from heterozygosity for an inhibitor gene. In(Jk) in analogy with the In(Lu) dominant inhibitor of Lutheran and other antigens.
- Immunized individuals with the Jk(a-b-) phenotype may produce anti-Jk3.
- Very weak expression of Jka and/or Jkb can be detected on In(Jk) red blood cells in adsorption/elution tests.

Kidd Antibodies and Clinical Significance

- Antibody subtypes and complement fixation: Anti-Jka and -Jkb are not common. They are usually warm-reacting IgG1 and IgG3 but may also include IgG2, IgG4 or IgM. Approximately 50% of anti-Jka and -Jkb antibodies are capable of binding complement.
- Dosage: Kidd antibodies display dosage: red cells from homozygous individuals (JkaJka or Jk(a+b-)) express more antigen than heterozygous individuals (JkaJkb or Jk(a+b+)). Anti-Kidd antibodies appear to react more strongly against cells that are homozygous.
- Laboratory detection: Kidd antibodies can be difficult to detect by direct agglutination testing and generally require addition of antihuman globulin (AHG).
- Clinical significance: Kidd antibodies are dangerous as they are capable of causing severe acute hemolytic transfusion reactions ('kidd kills').

They are unique in that they are capable of dropping to low or even undetectable levels after several months following an exposure event. Thus, on pre-transfusion testing, an anti-Jka or -Jkb may go undetected. Following transfusion, a subsequent robust antibody response in the patient can occur (anamnestic response), resulting in hemolysis of the transfused red blood cells. Kidd antibodies are often capable of binding complement and causing intravascular hemolysis. More often, however, Kidd antibodies cause acute extravascular hemolysis. They often cause delayed hemolytic transfusion reactions (DHTRs). Kidd antibodies only rarely cause hemolytic disease of the fetus and newborn.

Kidd Glycoprotein as Urea Transporter

Kidd antigens are located on a red blood cell urea transporter (aka human urea transporter 11- HUT11 or UT-B1). As red blood cells approach the renal medulla (where there is a high concentration of urea), the urea transporter allows for rapid uptake of urea and prevents cell shrinkage in the hypertonic environment of the medulla. As the red cell leaves the medulla, the urea is transported back out of the cell, preventing cellular swelling and preventing the urea from being carried away from the kidney. HUT11 was detected on endothelial cells of the vasa recta (vascular supply of the renal medulla) but it is not present in renal tubules.

Due to absence of the urea transporter, Jk(a-b-) cells are not hemolyzed by 2M urea. This can be used as a screening test for Jk(a-b-) donors. The Jk(a-b-) phenotype has no clinical defect, although two individuals with this phenotype have been reported to have mild urine-concentrating defects.

Kidd Antibodies in Transplant Patients

Kidd antibodies are capable of behaving as histocompatibility antigens in renal transplants and may be responsible for allograft rejection in some cases.

Rhag

Rh-associated glycoprotein (RHAG) is an ammonia transporter protein that in humans is encoded by the RHAG gene. RHAG has also recently been designated CD241 (cluster of differentiation 241). Mutations in the RHAG gene can cause stomatocytosis.

The Rh blood group antigens (MIM 111700) are associated with human erythrocyte membrane proteins of approximately 30 kD, the so-called Rh30 polypeptides. Heterogeneously glycosylated membrane proteins of 50 and 45 kD, the Rh50 glycoproteins, are coprecipitated with the Rh30 polypeptides on immunoprecipitation with anti-Rh-specific mono- and polyclonal antibodies. The Rh antigens appear to exist as a multisubunit complex of CD47 (MIM 601028),

LW (MIM 111250), glycophorin B (MIM 111740), and play a critical role in the Rh50 glycoprotein [supplied by OMIM].

Calcium ATPase

Ca^{2+} ATPase is a form of P-ATPase that transfers calcium after a muscle has contracted. The calcium ATPase are:

- Plasma membrane Ca^{2+} ATPase (PMCA)
- Sarcoplasmic reticulum Ca^{2+} ATPase (SERCA)

Plasma Membrane Ca^{2+} ATPase (PMCA)

The plasma membrane Ca^{2+} ATPase (PMCA) is a transport protein in the plasma membrane of cells that serves to remove calcium (Ca^{2+}) from the cell. It is vital for regulating the amount of Ca^{2+} within cells. In fact, the PMCA is involved in removing Ca^{2+} from all eukaryotic cells. There is a very large transmembrane electrochemical gradient of Ca^{2+} driving the entry of the ion into cells, yet it is very important for cells to maintain low concentrations of Ca^{2+} for proper cell signalling; thus it is necessary for the cell to employ ion pumps to remove the Ca^{2+}. The PMCA and the sodium calcium exchanger (NCX) are together the main regulators of intracellular Ca^{2+} concentrations. Since it transports Ca^{2+} into the extracellular space, the PMCA is also an important regulator of the calcium concentration in the extracellular space. The PMCA belongs to a family of P-type primary ion transport ATPases that form an aspartyl phosphate intermediate. The PMCA is expressed in a variety of tissues, including the brain.

Sarcoplasmic Reticulum Ca^{2+} ATPase (SERCA)

SERCA resides in the sarcoplasmic reticulum (SR) within muscle cells. It is a Ca^{2+} ATPase that transfers Ca^{2+} from the cytosol of the cell to the lumen of the SR at the expense of ATP hydrolysis during muscle relaxation.

Na-K-Cl Cotransporter

The Na-K-Cl cotransporter (NKCC) is a protein that aids in the active transport of sodium, potassium, and chloride into and out of cells. There are two varieties of this membrane transport protein, NKCC1 and NKCC2, however these are encoded by two different genes (SLC12A2 and SLC12A1 respectively) and are not isoforms. Two isoforms of the NKCC1/Slc12a2 gene result from keeping (isoform 1) or skipping (isoform 2) exon 21 in the final gene product.

NKCC1 is widely distributed throughout the body; it has important functions in organs that secrete fluids. NKCC2 is found specifically in the kidney, where it serves to extract sodium, potassium, and chloride from the urine so that they can be reabsorbed into the blood.

Function

NKCC proteins are membrane transport proteins that transport sodium (Na), potassium (K), and chloride (Cl) ions across the cell membrane. Because they move each solute in the same direction, NKCC proteins are considered symporters. They maintain electroneutrality by moving two positively charged solutes (sodium and potassium) alongside two parts of a negatively charged solute (chloride). Thus the stoichiometry of the NKCC proteins is 1Na:1K:2Cl.

NKCC1

NKCC1 is widely distributed throughout the body, especially in organs that secrete fluids, called exocrine glands. In cells of these organs, NKCC1 is commonly found in the basolateral membrane, the part of the cell membrane closest to the blood vessels. Its basolateral location gives NKCC1 the ability to transport sodium, potassium, and chloride from the blood into the cell. Other transporters assist in the movement of these solutes out of the cell through its apical surface. The end result is that solutes from the blood, particularly chloride, are secreted into the lumen of these exocrine glands, increasing the luminal concentration of solutes and causing water to be secreted by osmosis.

In addition to exocrine glands, NKCC1 is necessary for establishing the potassium-rich endolymph that bathes part of the cochlea, an organ necessary for hearing. Inhibition of NKCC1, as with furosemide or other loop diuretics, can result in deafness.

NKCC1 is also expressed in many regions of the brain during early development, but not in adulthood. This change in NKCC1 presence seems to be responsible for altering responses to the neurotransmitters GABA and glycine from excitatory to inhibitory, which was suggested to be important for early neuronal development. As long as NKCC1 transporters are predominantely active, internal chloride concentrations in neurons is raised in comparison with mature chloride concentrations, which is important for GABA and glycine responses, as respective ligand-gated anion channels are permeable to chloride. With higher internal chloride concentrations, outward driving force for this ions increases, and thus channel opening leads to chloride leaving the cell, thereby depolarizing it. Put another way, increasing internal chloride concentration increases the reversal potential for chloride, given by the Nernst equation. Later in development expression of NKCC1 is reduced, while expression of a KCC2 K-Cl cotransporter increased, thus bringing internal chloride concentration in neurons down to adult values.

NKCC2

NKCC2 is specifically found in cells of the thick ascending limb of the loop of Henle in nephrons, the basic functional units of the kidney. Within these cells, NKCC2 resides in the apical membrane abutting the nephron's lumen, the hollow space containing urine. Urine in the thick ascending limb of the loop of Henle has a relatively high concentration of sodium. That is, the electrochemical gradient of sodium favors movement of sodium from the urine and into cells. At this region of the nephron, NKCC2 is the major transport protein by which sodium is reabsorbed from the urine and into cells. According to the stoichiometry outlined above, each molecule of sodium reabsorbed brings one molecule of potassium and two molecules of chloride. Sodium goes on to be reabsorbed into the blood, where it contributes to the maintenance of blood pressure.

Furosemide and other loop diuretics inhibit the activity of NKCC2, thereby impairing sodium reabsorption in the thick ascending limb of the loop of Henle. Impaired sodium reabsorption prevents the thick ascending limb from contributing to maintenance of blood pressure. Loop diuretics therefore ultimately result in decreased blood pressure.

The hormone, Vasopressin, stimulates the activity of NKCC2. Vasopressin stimulates sodium chloride reabsorption in the thick ascending limb of the nephron activating signaling pathways. It increases the traffic of the NKCC2 to the membrane and phosphorylate at some serine and threonine sites the cytoplasmic N-terminal of the NKCC2 located in the membrane, increasing its activity; further aiding in water reabsorption in the collecting duct through Aquaporin 2 channels by the creation of a hypo-osmotic filtrate.

Genetics

NKCC1 and NKCC2 are encoded by genes on the long arms of chromosomes 15 and 5, respectively. A loss of function mutation of NKCC2 produces Bartter syndrome, an autosomal recessive disorder characterized by hypokalemic metabolic alkalosis with normal to low blood pressure.

Kinetics

The energy required to move solutes across the cell membrane is provided by the electrochemical gradient of sodium. Sodium's electrochemical gradient is established by the Na-K ATPase, which is an ATP-dependent enzyme. Since NKCC proteins use sodium's gradient, their activity is indirectly dependent on ATP; for this reason, NKCC proteins are said to move solutes by way of secondary active transport.

Sodium-Chloride Symporter

The sodium-chloride symporter (also known as Na^+-Cl" cotransporter, abbreviated as NCC or NCCT, or as the thiazide-sensitive Na^+-Cl" cotransporter or TSC for short) is a cotransporter in the kidney which has the function of reabsorbing sodium and chloride ions from the tubular fluid into the cells of the distal convoluted tubule of the nephron. It is a member of the SLC12 cotransporter family of electroneutral cation-coupled chloride cotransporters. In humans, it is encoded by the gene SLC12A3 (solute carrier family 12 member 3) located in 16q13.

Molecular Biology

The sodium-chloride symporter or NCC is a member of the SLC12 cotransporter family of electroneutral cation-coupled chloride cotransporter, along with the potassium-chloride cotransporters (K^+-Cl" cotransporters or KCCs), the sodium-potassium-chloride cotransporters (Na^+-K^+-Cl" cotransporters or NKCCs) and orphan member CIP (cotransporter interacting protein) and CCC9.

The sodium-chloride symporter's protein sequence has a high degree of identity between different mammalian species (over 90% between human, rat and mouse). The SLC12A3 gene encodes for a protein of 1,002 to 1,030 amino acid residues. NCC is a transmembrane protein, presumed to have a hydrophobic core of either 10 or 12 transmembrane domains with intracellular amino- and carboxyl-terminus domains. The exact structure of the NCC protein is unknown, as it has not yet been crystallized. The NCC protein forms homodimers at the plasma membrane.

N-glycosylation occurs in two sites in a long extracellular loop connecting two transmembrane domains within the hydrophobic core. This posttranslational modification is necessary for proper folding and transport of the protein to the plasma membrane.

Function

Because NCC is located at the apical membrane of the distal convoluted tubule of the nephron, it faces the lumen of the tubule and is in contact with the tubular fluid. Using the sodium gradient across the apical membrane of the cells in distal convoluted tubule, the sodium-chloride symporter transports Na^+ and Cl" from the tubular fluid into these cells. Afterward, the Na^+ is pumped out of the cell and into the bloodstream by the Na^+-K^+ ATPase located at the basal membrane and the Cl" leaves the cells through the basolateral chloride channel ClC-Kb. The sodium-chloride symporter accounts for the absorption of 5% of the salt filtered at the glomerulus. NCC activity is known to have two control mechanisms affecting protein trafficking to the plasma membrane and transporter kinetics by phosphorylation and de-phosphorylation of conserved serine/threonine residues.

As NCC has to be at the plasma membrane to function, its activity can be regulated by increasing or decreasing the amount of protein at the plasma membrane. Some NCC modulators, such as the WNK3 and WNK4 kinases may regulate the amount of NCC at the cell surface by inducing the insertion or removal, respectively, of the protein from the plasma membrane.

Furthermore, many residues of NCC can be phosphorylated or dephosphorylated to activate or inhibit NCC uptake of Na^+ and Cl". Other NCC modulators, including intracellular chloride depletion, angiotensin II, aldosterone and vasopressin, can regulate NCC activity by phosphorylating conserved serine/threonine residues. NCC activity can be inhibited by thiazides, which is why this symporter is also known as the thiazide-sensitive Na^+-Cl" cotransporter.

Pathology

Gitelman's Syndrome

A loss of NCC function is associated with Gitelman syndrome, an autosomic recessive disease characterized by salt wasting and low blood pressure, hypokalemic metabolic alkalosis, hypomagnesemia and hypocalciuria.

Over a hundred different mutations in the NCC gene have been described as causing Gitelman syndrome, including nonsense, frameshift, splice site and missense mutations. Two different types of mutations exist within the group of missense mutations causing loss of NCC function. Type I mutations cause a complete loss of NCC function, in which the synthesized protein is not properly glycosylated. NCC protein harboring type I mutations is retained in the endoplasmic reticulum and cannot be trafficked to the cell surface. Type II mutations cause a partial loss of NCC function in which the cotransporter is trafficked to the cell surface but has an impaired insertion in the plasma membrane. NCC harboring type II mutations have normal kinetic properties but are present in lower amounts at the cell surface, resulting in a decreased uptake of sodium and chloride. NCC harboring type II mutations is still under control of its modulators and can still increase or decrease its activity in response to stimuli, whereas type I mutations cause a complete loss of function and regulation of the cotransporter. However, in some patients with Gitelman's syndrome, no mutations in the NCC gene have been found despite extensive genetic work-up.

Hypertension and Blood Pressure

NCC has also been implicated to play a role in control of blood pressure in the open population, with both common polymorphisms and rare mutations altering NCC function, renal salt reabsorption and, presumably, blood pressure. Individuals with rare mutations in genes responsible for salt control in the kidney, including NCC, have been found to have a lower blood pressure than controls. NCC harboring these mutations has a lower function than wild-type cotransporter although some mutations found in individuals in the open population seem to be less deleterious to cotransporter function than mutations in individuals with Gitelman's syndrome.

Furthermore, heterozygous carriers of mutations causing Gitelman syndrome (i.e. individuals who have a mutation in one of the two alleles and do not have the disease) have a lower blood pressure than non-carriers in the same family.

Pseudohypoaldosteronism Type II

Type II pseudohypoaldosteronism (PHA2), also known as Gordon's syndrome, is an autosomal dominant disease in which there is an increase in NCC activity leading to short stature, increased blood pressure, increased serum K^+ levels, increased urinary calcium excretion and hyperchloremic metabolic acidosis. However, PHA2 is not caused by mutations within the NCC gene, but by mutations in NCC regulators WNK1 and WNK4. Patients respond well to treatment with thiazide-type diuretics.

Sodium–Hydrogen Antiporter

The sodium–hydrogen antiporter or sodium–hydrogen exchanger is a membrane protein found in many cells, and especially in those of the nephron of the kidney. Specifically in the intercalary cells of the collecting duct. The membrane pump is primarily responsible for maintaining the balance of pH.

Isoforms

There are several isoforms of the antiporter:

- Sodium–hydrogen antiporter 1
- Sodium–hydrogen antiporter 2
- Sodium–hydrogen antiporter 3
- Sodium–hydrogen antiporter 5
- Sodium–hydrogen antiporter 6
- Sodium–hydrogen antiporter 8

Chloride Potassium Symporter

The chloride potassium symporter is a membrane transport protein that is present in the S3-segment of the renal proximal tubule and in the neuron. It functions in renal chloride reabsorption to transport chloride across the basolateral membrane. They can lower intracellular chloride concentrations below the electrochemical equilibrium potential. They are of solute carrier family 12.

Types

Types of chloride potassium symporter are:

- Chloride potassium symporter 4
- Chloride potassium symporter 5

Each is encoded by a separate gene. The preceding symporters are actually other types of ion pumps. Member 3, for instance, is the sodium-chloride symporter.

KCNN4

Potassium intermediate/small conductance calcium-activated channel, subfamily N, member 4, also known as KCNN4, is a human gene encoding the $K_{Ca}3.1$ protein.

Function

The $K_{Ca}3.1$ protein is part of a potentially heterotetrameric voltage-independent potassium channel that is activated by intracellular calcium. Activation is followed by membrane depolarization, which promotes calcium influx. The encoded protein may be part of the predominant calcium-activated potassium channel in T-lymphocytes. This gene is similar to other KCNN family potassium channel genes, but it differs enough to possibly be considered as part of a new subfamily.

History

The channel activity was first described in 1958 by Gyorgi Gardos in human erythrocytes. The channels is also named Gardos channel because of its discoverer.

Na^+/K^+-ATPase

Na+/K+-ATPase (sodium-potassium adenosine triphosphatase, also known as the Na+/K+pump or sodium-potassium pump) is an enzyme (EC 3.6.3.9) (an electrogenic transmembrane ATPase) found in the plasma membrane of all animal cells. The Na+/K+-ATPase enzyme is a solute pump that pumps sodium out of cells while pumping potassium into cells, both against their concentration gradients. This pumping is active (it uses energy from ATP) and is important for cell physiology. An example application is nerve conduction. It has antiporter-like activity but is not actually an antiporter since both molecules are moving against their concentration gradient.

Sodium-Potassium Pumps

Active transport is responsible for the fact that cells contain a relatively high concentrations of potassium ions but low concentrations of sodium ions. The mechanism responsible for this is the sodium-potassium pump, which moves these two ions in opposite directions across the plasma membrane. This was investigated by following the passage of radioactively labeled ions across the plasma membrane of certain cells. It was found that the concentrations of sodium and potassium ions on the two sides of the membrane are interdependent, suggesting that the same carrier transports both ions. It is now known that the carrier is an ATP-ase and that it pumps three sodium ions out of the cell for every two potassium ions pumped in.

The sodium-potassium pump was discovered in the 1950s by a Danish scientist, Jens Christian Skou, who was awarded a Nobel Prize in 1997. It marked an important step forward in the understanding of how ions get into and out of cells, and it has a particular significance for excitable cells such as nervous cells, which depend on this pump for responding to stimuli and transmitting impulses.

Function

The Na+/K+-ATPase helps maintain resting potential, effect transport, and regulate cellular volume. It also functions as a signal transducer/integrator to regulate MAPK pathway, ROS, as well as intracellular calcium. In most animal cells, the Na+/K+-ATPase is responsible for about 1/5 of the cell's energy expenditure. For neurons, the Na+/K+-ATPase can be responsible for up to 2/3 of the cell's energy expenditure.

Resting Potential

In order to maintain the cell membrane potential, cells keep a low concentration of sodium ions and high levels of potassium ions within the cell (intracellular). The sodium-potassium pump moves 3 sodium ions out and moves 2 potassium ions in, thus, in total, removing one positive charge carrier from the intracellular space.

The action of the sodium-potassium pump is not the only mechanism responsible for the generation of the resting membrane potential. Also, the selective permeability of the cell's plasma membrane for the different ions plays an important role. All mechanisms involved are explained in the main article on generation of the resting membrane potential.

Transport

Export of sodium from the cell provides the driving force for several secondary active transporters membrane transport proteins, which import glucose, amino acids, and other nutrients into the cell by use of the sodium gradient.

Another important task of the Na+-K+pump is to provide a Na+gradient that is used by certain carrier processes. In the gut, for example, sodium is transported out of the reabsorbing cell on the blood (interstitial fluid) side via the Na+-K+pump, whereas, on the reabsorbing (lumenal) side, the Na+-glucose symporter uses the created Na+gradient as a source of energy to import both Na+and glucose, which is far more efficient than simple diffusion. Similar processes are located in the renal tubular system.

Controlling Cell Volume

Failure of the Na+-K+pumps can result in swelling of the cell. A cell's osmolarity is the sum of the concentrations of the various ion species and many proteins and other organic compounds inside the cell. When this is higher than the osmolarity outside of the cell, water flows into the cell through osmosis. This can cause the cell to swell up and lyse. The Na+-K+pump helps to maintain the right concentrations of ions. Furthermore, when the cell begins to swell, this automatically activates the Na+-K+pump.

Functioning as Signal Transducer

Within the last decade [when?], many independent labs have demonstrated that, in addition to the classical ion transporting, this membrane protein can also relay extracellular ouabain-binding signalling into the cell through regulation of protein tyrosine phosphorylation. The downstream signals through ouabain-triggered protein phosphorylation events include activation of the mitogen-activated protein kinase (MAPK) signal cascades, mitochondrial reactive oxygen species (ROS) production, as well as activation of phospholipase C (PLC) and inositol triphosphate (IP3) receptor (IP3R) in different intracellular compartments.

Protein-protein interactions play a very important role in Na+ -K+ pump-mediated signal transduction. For example, Na+ -K+ pump interacts directly with Src, a non-receptor tyrosine kinase, to form a signaling receptor complex. Src kinase is inhibited by Na+ -K+ pump, while, upon ouabain binding, the Src kinase domain will be released and then activated. Based on this scenario, NaKtide, a peptide Src inhibitor derived from Na+ -K+ pump, was developed as a functional ouabain-Na+ -K+ pump-mediated signal transduction. Na+ -K+ pump also interacts with ankyrin, IP3R, PI3K, PLC-gamma and cofilin.

Controlling Neuron Activity States

The Na+-K+pump has been shown to control and set the intrinsic activity mode of cerebellar Purkinje neurons. This suggests that the pump might not simply be a homeostatic, "housekeeping" molecule for ionic gradients; but could be a computation element in the cerebellum and the brain. Indeed, a mutation in the Na+-K+pump causes rapid onset dystonia parkinsonism, which has symptoms to indicate that it is a pathology of cerebellar computation. Furthermore, an ouabain block of Na+-K+ pumps in the cerebellum of a live mouse results in it displaying ataxia and dystonia.

Alcohol inhibits sodium-potassium pumps in the cerebellum and this is likely how it corrupts cerebellar computation and body co-ordination. The distribution of the Na+-K+ pump on myelinated axons, in human brain, was demonstrated to be along the internodal axolemma, and not within the nodal axolemma as previously thought.

Mechanism

- The pump, after binding ATP, binds 3 intracellular Na+ions.
- ATP is hydrolyzed, leading to phosphorylation of the pump at a highly conserved aspartate residue and subsequent release of ADP.
- A conformational change in the pump exposes the Na+ions to the outside. The phosphorylated form of the pump has a low affinity for Na+ions, so they are released.
- The pump binds 2 extracellular K+ions. This causes the dephosphorylation of the pump, reverting it to its previous conformational state, transporting the K+ions into the cell.
- The unphosphorylated form of the pump has a higher affinity for Na+ions than K+ions, so the two bound K+ions are released. ATP binds, and the process starts again.

Regulation

Endogenous

The Na+/K+-ATPase is upregulated by cAMP. Thus, substances causing an increase in cAMP upregulate the Na+/K+-ATPase. These include the ligands of the G_s-coupled GPCRs. In contrast, substances causing a decrease in cAMP downregulate the Na+/K+-ATPase. These include the ligands of the G_i-coupled GPCRs. Note: Early studies indicated the opposite effect, but these were later found to be inaccurate due to additional complicating factors.

Exogenous

The Na+ -K+-ATPase can be pharmacologically modified by administrating drugs exogenously. For instance, Na+-K+-ATPase found in the membrane of heart cells is an important target of cardiac glycosides (for example digoxin and ouabain), inotropic drugs used to improve heart performance by increasing its force of contraction.

Muscle contraction is dependent on a 100- to 10,000-times-higher-than-resting intracellular Ca2+concentration, which is caused by Ca2+release from the muscle cells' sarcoplasmic reticulum. Immediately after muscle contraction, intracellular Ca2+is quickly returned to its normal concentration by a carrier enzyme in the plasma membrane, and a calcium pump in sarcoplasmic reticulum, causing the muscle to relax.

Since this carrier enzyme (Na+-Ca2+translocator) uses the Na gradient generated by the Na+-K+pump to remove Ca2+from the intracellular space, slowing down the Na+-K+pump results in a permanently elevated Ca2+ level in the muscle, which may be the mechanism of the long-term inotropic effect of cardiac glycosides such as digoxin.

Discovery

Na+/K+-ATPase was discovered by Jens Christian Skou in 1957 while working as assistant professor at the Department of Physiology, University of Aarhus, Denmark. He published his work that year. In 1997, he received one-half of the Nobel Prize in Chemistry "for the first discovery of an ion-transporting enzyme, Na+, K+-ATPase."

CELL ADHESION

- ICAM-4 – interacts with integrins;
- BCAM – a glycoprotein that defines the Lutheran blood group and also known as Lu or laminin-binding protein.

Structural role – The following membrane proteins establish linkages with skeletal proteins and may play an important role in regulating cohesion between the lipid bilayer and membrane skeleton, likely enabling the red cell to maintain its favorable membrane surface area by preventing the membrane from collapsing (vesiculating).

- Ankyrin-based macromolecular complex – proteins linking the bilayer to the membrane skeleton through the interaction of their cytoplasmic domains with Ankyrin.
  - Band 3 – also assembles various glycolytic enzymes, the presumptive CO_2 transporter, and carbonic anhydrase into a macromolecular complex termed a "metabolon," which may play a key role in regulating red cell metabolism and ion and gas transport function);
  - RhAG – also involved in transport, defines associated unusual blood group phenotype Rh_{mod}.
- Protein 4.1R-based macromolecular complex – proteins interacting with Protein 4.1R.
  - Protein 4.1R – weak expression of Gerbich antigens;
  - Glycophorin C and D – glycoprotein, defines Gerbich Blood Group;
  - XK – defines the Kell Blood Group and the Mcleod unusual phenotype (lack of Kx antigen and greatly reduced expression of Kell antigens);
  - RhD/RhCE – defines Rh Blood Group and the associated unusual blood group phenotype Rh_{null};
  - Duffy protein – has been proposed to be associated with chemokine clearance;
  - Adducin – interaction with band 3;
  - Dematin- interaction with the Glut1 glucose transporter.

SURFACE ELECTROSTATIC POTENTIAL

The zeta potential is an electrochemical property of cell surfaces that is determined by the net electrical charge of molecules exposed at the surface of cell membranes of the cell. The normal zeta potential of the erythrocyte is “15.7 millivolts (mV). Much of this potential appears to be contributed by the exposed sialic acid residues in the membrane: their removal results in zeta potential of “6.06 mV.

CLINICAL NOTES

SEPARATION AND BLOOD DOPING

Red blood cells can be obtained from whole blood by centrifugation, which separates the cells from the blood plasma in a process known as blood fractionation. Packed red blood cells, which are made in this way from whole blood with the plasma removed, are used in transfusion medicine. During plasma donation, the red blood cells are pumped back into the body right away and only the plasma is collected.

Some athletes have tried to improve their performance by blood doping: first about 1 litre of their blood is extracted, then the red blood cells are isolated, frozen and stored, to be reinjected shortly before the competition. (Red blood cells can be conserved for 5 weeks at “79 °C or “110 °F) This practice is hard to detect but may endanger the human cardiovascular system which is not equipped to deal with blood of the resulting higher viscosity. Another method of blood doping involves injection with erythropoietin in order to stimulate production of red blood cells. Both practices are banned by the World Anti-Doping Agency.

ARTIFICIALLY GROWN RED BLOOD CELLS

In 2008 it was reported that human embryonic stem cells had been successfully coaxed into becoming erythrocytes in the lab. The difficult step was to induce the cells to eject their nucleus; this was achieved by growing the cells on stromal cells from the bone marrow. It is hoped that these artificial erythrocytes can eventually be used for blood transfusions.

DISEASES AND DIAGNOSTIC TOOLS

Blood diseases involving the red blood cells include:

- Anemias (or anaemias) are diseases characterized by low oxygen transport capacity of the blood, because of low red cell count or some abnormality of the red blood cells or the hemoglobin.
- Iron deficiency anemia is the most common anemia; it occurs when the dietary intake or absorption of iron is insufficient, and hemoglobin, which contains iron, cannot be formed

- Sickle-cell disease is a genetic disease that results in abnormal hemoglobin molecules. When these release their oxygen load in the tissues, they become insoluble, leading to mis-shaped red blood cells. These sickle shaped red cells are less deformable and viscoelastic meaning that they have become rigid and can cause blood vessel blockage, pain, strokes, and other tissue damage.
- Thalassemia is a genetic disease that results in the production of an abnormal ratio of hemoglobin subunits.
- Hereditary spherocytosis syndromes are a group of inherited disorders characterized by defects in the red blood cell's cell membrane, causing the cells to be small, sphere-shaped, and fragile instead of donut-shaped and flexible. These abnormal red blood cells are destroyed by the spleen. Several other hereditary disorders of the red blood cell membrane are known.
- Pernicious anemia is an autoimmune disease wherein the body lacks intrinsic factor, required to absorb vitamin B_{12} from food. Vitamin B_{12} is needed for the production of hemoglobin.
- Aplastic anemia is caused by the inability of the bone marrow to produce blood cells.
- Pure red cell aplasia is caused by the inability of the bone marrow to produce only red blood cells.
- Hemolysis is the general term for excessive breakdown of red blood cells. It can have several causes and can result in hemolytic anemia.
- The malaria parasite spends part of its life-cycle in red blood cells, feeds on their hemoglobin and then breaks them apart, causing fever. Both sickle-cell disease and thalassemia are more common in malaria areas, because these mutations convey some protection against the parasite.
- Polycythemias (or erythrocytoses) are diseases characterized by a surplus of red blood cells. The increased viscosity of the blood can cause a number of symptoms.
- In polycythemia vera the increased number of red blood cells results from an abnormality in the bone marrow.
- Several microangiopathic diseases, including disseminated intravascular coagulation and thrombotic microangiopathies, present with pathognomonic (diagnostic) red blood cell fragments called schistocytes. These pathologies generate fibrin strands that sever red blood cells as they try to move past a thrombus.
- Hemolytic transfusion reaction is the destruction of donated red blood cells after a transfusion, mediated by host antibodies, often as a result of a blood type mismatch.

Several blood tests involve red blood cells, including the RBC count (the number of red blood cells per volume of blood), the hematocrit (percentage of blood volume occupied by red blood cells), and the erythrocyte sedimentation rate. Many diseases involving red blood cells are diagnosed with a blood film (or peripheral blood smear), where a thin layer of blood is smeared on a microscope slide. The blood type needs to be determined to prepare for a blood transfusion or an organ transplantation.

3

Hemoglobin

Hemoglobin is the iron-containing oxygen-transport metalloprotein in the red blood cells of all vertebrates (with the exception of the fish family Channichthyidae) as well as the tissues of some invertebrates. Hemoglobin in the blood carries oxygen from the respiratory organs (lungs or gills) to the rest of the body (i.e. the tissues). There it releases the oxygen to permit aerobic respiration to provide energy to power the functions of the organism in the process called metabolism. In mammals, the protein makes up about 96% of the red blood cells' dry content (by weight), and around 35% of the total content (including water). Hemoglobin has an oxygen-binding capacity of 1.34 mL O_2 per gram, which increases the total blood oxygen capacity seventy-fold compared to dissolved oxygen in blood. The mammalian hemoglobin molecule can bind (carry) up to four oxygen molecules. Hemoglobin is involved in the transport of other gases: It carries some of the body's respiratory carbon dioxide (about 10% of the total) as carbaminohemoglobin, in which CO_2 is bound to the globin protein. The molecule also carries the important regulatory molecule nitric oxide bound to a globin protein thiol group, releasing it at the same time as oxygen. Hemoglobin is also found outside red blood cells and their progenitor lines. Other cells that contain hemoglobin include the A9 dopaminergic neurons in the substantia nigra, macrophages, alveolar cells, and mesangial cells in the kidney. In these tissues, hemoglobin has a non-oxygen-carrying function as an antioxidant and a regulator of iron metabolism. Hemoglobin and hemoglobin-like molecules are also found in many invertebrates, fungi, and plants. In these organisms, hemoglobins may carry oxygen, or they may act to transport and regulate other things such as carbon dioxide, nitric oxide, hydrogen sulfide and sulfide. A variant of the molecule, called leghemoglobin, is used to scavenge oxygen away from anaerobic systems, such as the nitrogen-fixing nodules of leguminous plants, before the oxygen can poison the system.

RESEARCH HISTORY

In 1825 J.F. Engelhard discovered that the ratio of Fe to protein is identical in the hemoglobins of several species. From the known atomic mass of iron he calculated the molecular mass of hemoglobin to n × 16000 (n = number of irons per hemoglobin, now known to be 4), the first determination of a protein's molecular mass. This "hasty conclusion" drew a lot of ridicule at the time from scientists who could not believe that any molecule could be that big. Gilbert Smithson Adair confirmed Engelhard's results in 1925 by measuring the osmotic pressure of hemoglobin solutions.

The oxygen-carrying protein hemoglobin was discovered by Hünefeld in 1840. In 1851, German physiologist Otto Funke published a series of articles in which he described growing hemoglobin crystals by successively diluting red blood cells with a solvent such as pure water, alcohol or ether, followed by slow evaporation of the solvent from the resulting protein solution. Hemoglobin's reversible oxygenation was described a few years later by Felix Hoppe-Seyler.

In 1959, Max Perutz determined the molecular structure of myoglobin (similar to hemoglobin) by X-ray crystallography. This work resulted in his sharing with John Kendrew the 1962 Nobel Prize in Chemistry.

The role of hemoglobin in the blood was elucidated by French physiologist Claude Bernard. The name hemoglobin is derived from the words heme and globin, reflecting the fact that each subunit of hemoglobin is a globular protein with an embedded heme group. Each heme group contains one iron atom, that can bind one oxygen molecule through ion-induced dipole forces. The most common type of hemoglobin in mammals contains four such subunits.

GENETICS

Hemoglobin consists mostly of protein subunits (the "globin" molecules), and these proteins, in turn, are folded chains of a large number of different amino acids called polypeptides. The amino acid sequence of any polypeptide created by a cell is in turn determined by the stretches of DNA called genes. In all proteins, it is the amino acid sequence that determines the protein's chemical properties and function.

There is more than one hemoglobin gene. The amino acid sequences of the globin proteins in hemoglobins usually differ between species. These differences grow with evolutionary distance between species. For example, the most common hemoglobin sequences in humans and chimpanzees are nearly identical, differing by only one amino acid in both the alpha and the beta globin protein chains. These differences grow larger between less closely related species.

Even within a species, different variants of hemoglobin always exist, although one sequence is usually a "most common" one in each species. Mutations in the genes for the hemoglobin protein in a species result in hemoglobin variants. Many of these mutant forms of hemoglobin cause no disease. Some of these mutant forms of hemoglobin, however, cause a group of hereditary diseases termed the hemoglobinopathies. The best known hemoglobinopathy is sickle-cell disease, which was the first human disease whose mechanism was understood at the molecular level. A (mostly) separate set of diseases called thalassemias involves underproduction of normal and sometimes abnormal hemoglobins, through problems and mutations in globin gene regulation. All these diseases produce anemia.

Variations in hemoglobin amino acid sequences, as with other proteins, may be adaptive. For example, recent studies have suggested genetic variants in deer mice that help explain how deer mice that live in the mountains are able to survive in the thin air that accompanies high altitudes. A researcher from the University of Nebraska-Lincoln found mutations in four different genes that can account for differences between deer mice that live in lowland prairies versus the mountains.

After examining wild mice captured from both highlands and lowlands, it was found that: the genes of the two breeds are "virtually identical–except for those that govern the oxygen-carrying capacity of their hemoglobin". "The genetic difference enables highland mice to make more efficient use of their oxygen", since less is available at higher altitudes, such as those in the mountains. Mammoth hemoglobin featured mutations that allowed for oxygen delivery at lower temperatures, thus enabling mammoths to migrate to higher latitudes during the Pleistocene.

SYNTHESIS

Hemoglobin (Hb) is synthesized in a complex series of steps. The heme part is synthesized in a series of steps in the mitochondria and the cytosol of immature red blood cells, while the globin protein parts are synthesized by ribosomes in the cytosol. Production of Hb continues in the cell throughout its early development from the proerythroblast to the reticulocyte in the bone marrow. At this point, the nucleus is lost in mammalian red blood cells, but not in birds and many other species. Even after the loss of the nucleus in mammals, residual ribosomal RNA allows further synthesis of Hb until the reticulocyte loses its RNA soon after entering the vasculature (this hemoglobin-synthetic RNA in fact gives the reticulocyte its reticulated appearance and name).

STRUCTURE

Hemoglobin has a quaternary structure characteristic of many multi-subunit globular proteins. Most of the amino acids in hemoglobin form alpha helices, connected by short non-helical segments. Hydrogen bonds stabilize the helical sections inside this protein, causing attractions within the molecule, folding each polypeptide chain into a specific shape. Hemoglobin's quaternary structure comes from its four subunits in roughly a tetrahedral arrangement.

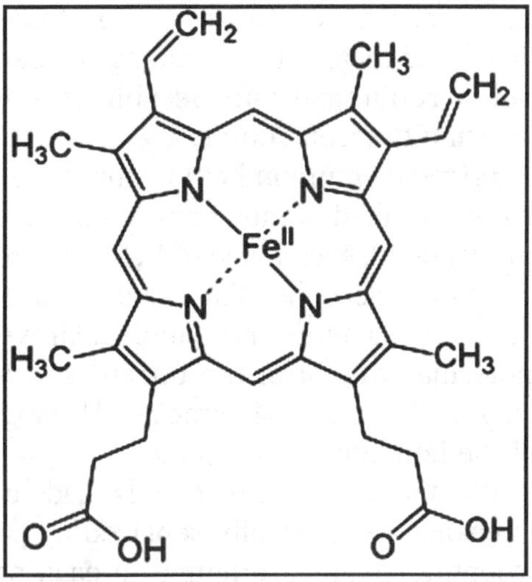

Fig. Heme b Group.

In most vertebrates, the hemoglobin molecule is an assembly of four globular protein subunits. Each subunit is composed of a protein chain tightly associated with a non-protein heme group. Each protein chain arranges into a set of alpha-helix structural segments connected together in a globin fold arrangement, so called because this arrangement is the same folding motif used in other heme/globin proteins such as myoglobin. This folding pattern contains a pocket that strongly binds the heme group.

A heme group consists of an iron (Fe) ion (charged atom) held in a heterocyclic ring, known as a porphyrin. This porphyrin ring consists of four pyrrole molecules cyclically linked together (by methine bridges) with the iron ion bound in the center. The iron ion, which is the site of oxygen binding, coordinates with the four nitrogens in the center of the ring, which all lie in one plane. The iron is bound strongly (covalently) to the globular protein via the imidazole ring of F8 histidine residue (also known as the proximal histidine) below the porphyrin ring. A sixth position can reversibly bind oxygen by a coordinate covalent bond, completing the octahedral group of six ligands. Oxygen binds in an "end-on bent" geometry where one oxygen atom binds Fe and the other protrudes at an angle. When oxygen is not bound, a very weakly bonded water molecule fills the site, forming a distorted octahedron.

Even though carbon dioxide is carried by hemoglobin, it does not compete with oxygen for the iron-binding positions but is bound to the protein chains of the structure.

The iron ion may be either in the Fe^{2+} or in the Fe^{3+} state, but ferrihemoglobin (methemoglobin) (Fe^{3+}) cannot bind oxygen. In binding, oxygen temporarily and reversibly oxidizes (Fe^{2+}) to (Fe^{3+}) while oxygen temporarily turns into superoxide, thus iron must exist in the +2 oxidation state to bind oxygen. If superoxide ion associated to Fe^{3+} is protonated, the hemoglobin iron will remain oxidized and incapable of binding oxygen. In such cases, the enzyme methemoglobin reductase will be able to eventually reactivate methemoglobin by reducing the iron center.

In adult humans, the most common hemoglobin type is a tetramer (which contains 4 subunit proteins) called hemoglobin A, consisting of two α and two β subunits non-covalently bound, each made of 141 and 146 amino acid residues, respectively. This is denoted as $\alpha_2\beta_2$. The subunits are structurally similar and about the same size. Each subunit has a molecular weight of about 16,000 daltons, for a total molecular weight of the tetramer of about 64,000 daltons (64,458 g/mol). Thus, 1 g/dL = 0.1551 mmol/L. Hemoglobin A is the most intensively studied of the hemoglobin molecules.

In human infants, the hemoglobin molecule is made up of 2 α chains and 2 γ chains. The gamma chains are gradually replaced by β chains as the infant grows. The four polypeptide chains are bound to each other by salt bridges, hydrogen bonds, and the hydrophobic effect.

OXYGEN SATURATION

In general, hemoglobin can be saturated with oxygen molecules (oxyhemoglobin), or desaturated with oxygen molecules (deoxyhemoglobin).

OXYHEMOGLOBIN

Oxyhemoglobin is formed during physiological respiration when oxygen binds to the heme component of the protein hemoglobin in red blood cells. This process occurs in the pulmonary capillaries adjacent to the alveoli of the lungs. The oxygen then travels through the blood stream to be dropped off at cells where it is utilized as a terminal electron acceptor in the production of ATP by the process of oxidative phosphorylation. It does not, however, help to counteract a decrease in blood pH. Ventilation, or breathing, may reverse this condition by removal of carbon dioxide, thus causing a shift up in pH.

Hemoglobin exists in two forms, a taut (tense) form (T) and a relaxed form (R). Various factors such as low pH, high CO_2 and high 2,3 BPG at the level of the tissues favor the taut form, which has low oxygen affinity and releases oxygen in the tissues. Conversely, a high pH, low CO_2, or low 2,3 BPG favors the relaxed form, which can better bind oxygen. The partial pressure of the system also affects O_2 affinity where, at high partial pressures of oxygen (such as those present in the alveoli), the relaxed (high affinity, R) state is favoured. Inversely, at low partial pressures (such as those present in respiring tissues), the (low affinity, T) tense state is favoured. Additionally, the binding of oxygen to the Iron-II heme pulls the iron into the plane of the porphyrin ring, causing a slight conformational shift. The shift encourages oxygen to bind to the three remaining hemes within hemoglobin (thus, oxygen binding is cooperative).

DEOXYGENATED HEMOGLOBIN

Deoxygenated hemoglobin is the form of hemoglobin without the bound oxygen. The absorption spectra of oxyhemoglobin and deoxyhemoglobin differ. The oxyhemoglobin has significantly lower absorption of the 660 nm wavelength than deoxyhemoglobin, while at 940 nm its absorption is slightly higher. This difference is used for measurement of the amount of oxygen in patient's blood by an instrument called pulse oximeter. This difference also accounts for the presentation of cyanosis, the blue to purplish color that tissues develop during hypoxia.

IRON'S OXIDATION STATE IN OXYHEMOGLOBIN

Deleting oxygenated hemoglobin's oxidation state is difficult because oxyhemoglobin (Hb-O_2), by experimental measurement, is diamagnetic (no net unpaired electrons), yet the low-energy electron configurations in both oxygen and iron are paramagnetic (suggesting at least one unpaired electron in the complex). The lowest-energy form of oxygen, and the lowest energy forms of the relevant oxidation states of iron, are these:

- Triplet oxygen, the highest-energy molecular oxygen species, has two unpaired electrons in antibonding π* molecular orbitals.
- Iron(II) tends to exist in a high-spin configuration where unpaired electrons exist in E_g antibonding orbitals.
- Iron(III) has an odd number of electrons, and thus must have one or more unpaired electrons, in any energy state.

All of these structures are paramagnetic (have unpaired electrons), not diamagnetic. Thus, a non-intuitive (e.g., a higher-energy for at least one species) distribution of electrons in the combination of iron and oxygen must exist, in order to explain the observed diamagnetism and no unpaired electrons.

The three logical possibilities to produce diamagnetic (no net spin) Hb-O_2 are:

1. Low-spin Fe^{2+} binds to singlet oxygen. Both low-spin iron and singlet oxygen are diamagnetic. However, the singlet form of oxygen is the higher-energy form of the molecule.
2. Low-spin Fe^{3+} binds to.$O_2^{''}$ (the superoxide ion) and the two unpaired electrons couple antiferromagnetically, giving diamagnetic properties.
3. Low-spin Fe^{4+} binds to peroxide, $O_2^{2''}$. Both are diamagnetic.

DIRECT EXPERIMENTAL DATA

- X-ray photoelectron spectroscopy suggests iron has an oxidation state of approximately 3.2
- Infrared vibrational frequencies of the O-O bond suggests a bond

length fitting with superoxide (a bond order of about 1.6, with superoxide being 1.5).

- X-ray Absorption Near Edge Structures at the iron K-edge. The energy shift of 5 eV between Deoxyhemoglobin and Oxyhemoglobin, as for all the Methemoglobin species, strongly suggests an actual local charge closer to Fe^{3+} than Fe^{2+}.

Thus, the nearest formal oxidation state of iron in Hb-O_2 is the +3 state, with oxygen in the "1 state (as superoxide.O_2"). The diamagnetism in this configuration arises from the single unpaired electron on superoxide aligning antiferromagnetically from the single unpaired electron on iron, to give no net spin to the entire configuration, in accordance with diamagnetic oxyhemoglobin from experiment.

The second choice of the three logical possibilities above for diamagnetic oxyhemoglobin being found correct by experiment, is not surprising: singlet oxygen (possibility #1) and large separations of charge (possibility #3) are both unfavorably high-energy states. Iron's shift to a higher oxidation state in Hb-O_2 decreases the atom's size, and allows it into the plane of the porphyrin ring, pulling on the coordinated histidine residue and initiating the allosteric changes seen in the globulins.

Early postulates by bio-inorganic chemists claimed that possibility #1 (above) was correct and that iron should exist in oxidation state II. This conclusion seemed likely, since the iron oxidation state III as methemoglobin, when not accompanied by superoxide.O_2" to "hold" the oxidation electron, was known to render hemoglobin incapable of binding normal triplet O_2 as it occurs in the air. It was thus assumed that iron remained as Fe(II) when oxygen gas was bound in the lungs. The iron chemistry in this previous classical model was elegant, but the required presence of the required diamagnetic high-energy singlet oxygen was never explained. It was classically argued that the binding of an oxygen molecule placed high-spin iron(II) in an octahedral field of strong-field ligands; this change in field would increase the crystal field splitting energy, causing iron's electrons to pair into the low-spin configuration, which would be diamagnetic in Fe(II). This forced low-spin pairing is indeed thought to happen in iron when oxygen binds, but is not enough to explain iron's change in size. Extraction of an additional electron from iron by oxygen is required to explain both iron's smaller size and observed increased oxidation state, and oxygen's weaker bond.

The assignment of a whole-number oxidation state is a formalism, as the covalent bonds are not required to have perfect bond orders involving whole electron transfer. Thus, all three models for paramagnetic Hb-O_2 may contribute to some small degree (by resonance) to the actual electronic configuration of Hb-O_2. However, the model of iron in Hb-O_2 being Fe(III) is more correct than the classical idea that it remains Fe(II).

COOPERATIVITY

When oxygen binds to the iron complex, it causes the iron atom to move back toward the center of the plane of the porphyrin ring. At the same time, the imidazole side-chain of the histidine residue interacting at the other pole of the iron is pulled toward the porphyrin ring. This interaction forces the plane of the ring sideways toward the outside of the tetramer, and also induces a strain in the protein helix containing the histidine as it moves nearer to the iron atom. This strain is transmitted to the remaining three monomers in the tetramer, where it induces a similar conformational change in the other heme sites such that binding of oxygen to these sites becomes easier.

In the tetrameric form of normal adult hemoglobin, the binding of oxygen is, thus, a cooperative process. The binding affinity of hemoglobin for oxygen is increased by the oxygen saturation of the molecule, with the first oxygens bound influencing the shape of the binding sites for the next oxygens, in a way favorable for binding. This positive cooperative binding is achieved through steric conformational changes of the hemoglobin protein complex as discussed above; i.e., when one subunit protein in hemoglobin becomes oxygenated, a conformational or structural change in the whole complex is initiated, causing the other subunits to gain an increased affinity for oxygen. As a consequence, the oxygen binding curve of hemoglobin is sigmoidal, or S-shaped, as opposed to the normal hyperbolic curve associated with noncooperative binding. The dynamic mechanism of the cooperativity in hemoglobin and its relation with the low-frequency resonance has been discussed.

BINDING FOR LIGANDS OTHER THAN OXYGEN

Besides the oxygen ligand, which binds to hemoglobin in a cooperative manner, hemoglobin ligands also include competitive inhibitors such as carbon monoxide (CO) and allosteric ligands such as carbon dioxide (CO_2) and nitric oxide (NO). The carbon dioxide is bound to amino groups of the globin proteins as carbaminohemoglobin, and is thought to account for about 10% of carbon dioxide transport in mammals. Nitric oxide is bound to specific thiol groups in the globin protein to form an S-nitrosothiol, which dissociates into free nitric oxide and thiol again, as the hemoglobin releases oxygen from its heme site. This nitric oxide transport to peripheral tissues is hypothesized to assist oxygen transport in tissues, by releasing vasodilatory nitric oxide to tissues in which oxygen levels are low.

COMPETITIVE

The binding of oxygen is affected by molecules such as carbon monoxide (CO) (for example, from tobacco smoking, car exhaust, and incomplete combustion in furnaces). CO competes with oxygen at the heme binding site. Hemoglobin binding affinity for CO is 250 times greater than its affinity for oxygen, meaning that small amounts of CO dramatically reduce hemoglobin's ability to transport oxygen. Since carbon monoxide is a colorless, odorless and tasteless gas, and poses a potentially fatal threat, detectors have become commercially available to warn of dangerous levels in residences. When hemoglobin combines with CO, it forms a very bright red compound called carboxyhemoglobin, which may cause the skin of CO poisoning victims to appear pink in death, instead of white or blue. When inspired air contains CO levels as low as 0.02%, headache and nausea occur; if the CO concentration is increased to 0.1%, unconsciousness will follow. In heavy smokers, up to 20% of the oxygen-active sites can be blocked by CO.

In similar fashion, hemoglobin also has competitive binding affinity for cyanide (CN^-), sulfur monoxide (SO), nitric oxide (NO), and sulfide (S^{2-}), including hydrogen sulfide (H_2S). All of these bind to iron in heme without changing its oxidation state, but they nevertheless inhibit oxygen-binding, causing grave toxicity.

The iron atom in the heme group must initially be in the ferrous (Fe^{2+}) oxidation state to support oxygen and other gases' binding and transport (it temporarily switches to ferric during the time oxygen is bound, as explained above). Initial oxidation to the ferric (Fe^{3+}) state without oxygen converts hemoglobin into "hemiglobin" or methemoglobin (pronounced "MET-hemoglobin"), which cannot bind oxygen. Hemoglobin in normal red blood cells is protected by a reduction system to keep this from happening. Nitric oxide is capable of converting a small fraction of hemoglobin to methemoglobin in red blood cells. The latter reaction is a remnant activity of the more ancient nitric oxide dioxygenase function of globins.

ALLOSTERIC

Carbon dioxide occupies a different binding site on the hemoglobin. Carbon dioxide is more readily dissolved in deoxygenated blood, facilitating its removal from the body after the oxygen has been released to tissues undergoing metabolism. This increased affinity for carbon dioxide by the venous blood is known as the Haldane effect. Through the enzyme carbonic anhydrase, carbon dioxide reacts with water to give carbonic acid, which decomposes into bicarbonate and protons:

$$CO_2 + H_2O \rightarrow H_2CO_3 \rightarrow HCO_3^- + H^+$$

Hence, blood with high carbon dioxide levels is also lower in pH (more acidic). Hemoglobin can bind protons and carbon dioxide, which causes a conformational change in the protein and facilitates the release of oxygen. Protons bind at various places on the protein, while carbon dioxide binds at the á-amino group. Carbon dioxide binds to hemoglobin and forms carbaminohemoglobin. This decrease in hemoglobin's affinity for oxygen by the binding of carbon dioxide and acid is known as the Bohr effect (shifts the O_2-

saturation curve to the right). Conversely, when the carbon dioxide levels in the blood decrease (i.e., in the lung capillaries), carbon dioxide and protons are released from hemoglobin, increasing the oxygen affinity of the protein. A reduction in the total binding capacity of hemoglobin to oxygen (i.e. shifting the curve down, not just to the right) due to reduced pH is called the root effect. This is seen in bony fish.

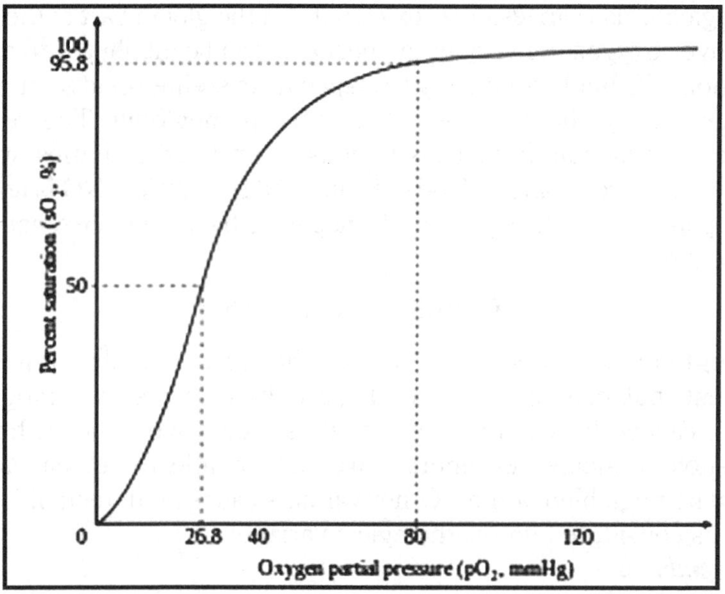

Fig. The Sigmoidal Shape of Hemoglobin's Oxygen-Dissociation Curve Results from Cooperative Binding of Oxygen to Hemoglobin.

It is necessary for hemoglobin to release the oxygen that it binds; if not, there is no point in binding it. The sigmoidal curve of hemoglobin makes it efficient in binding (taking up O_2 in lungs), and efficient in unloading (unloading O_2 in tissues).

In people acclimated to high altitudes, the concentration of 2,3-Bisphosphoglycerate (2,3-BPG) in the blood is increased, which allows these individuals to deliver a larger amount of oxygen to tissues under conditions of lower oxygen tension. This phenomenon, where molecule Y affects the binding of molecule X to a transport molecule Z, is called a heterotropic allosteric effect.

Animals other than humans use different molecules to bind to hemoglobin and change its O_2 affinity under unfavorable conditions. Fish use both ATP and GTP. These bind to a phosphate "pocket" on the fish hemoglobin molecule, which stabilizes the tense state and therefore decreases oxygen affinity. GTP reduces hemoglobin oxygen affinity much more than ATP, which is thought to be due to an extra hydrogen bond formed that further stabilizes the tense state. Under hypoxic conditions, the concentration of both ATP and GTP is reduced in fish red blood cells to increase oxygen affinity.

A variant hemoglobin, called fetal hemoglobin (HbF, $\alpha_2\gamma_2$), is found in the developing fetus, and binds oxygen with greater affinity than adult hemoglobin. This means that the oxygen binding curve for fetal hemoglobin is left-shifted (i.e., a higher percentage of hemoglobin has oxygen bound to it at lower oxygen tension), in comparison to that of adult hemoglobin. As a result, fetal blood in the placenta is able to take oxygen from maternal blood.

Hemoglobin also carries nitric oxide (NO) in the globin part of the molecule. This improves oxygen delivery in the periphery and contributes to the control of respiration. NO binds reversibly to a specific cysteine residue in globin; the binding depends on the state (R or T) of the hemoglobin. The resulting S-nitrosylated hemoglobin influences various NO-related activities such as the control of vascular resistance, blood pressure and respiration. NO is not released in the cytoplasm of erythrocytes but transported by an anion exchanger called AE1 out of them.

TYPES IN HUMANS

Hemoglobin variants are a part of the normal embryonic and fetal development, but may also be pathologic mutant forms of hemoglobin in a population, caused by variations in genetics. Some well-known hemoglobin variants such as sickle-cell anemia are responsible for diseases, and are considered hemoglobinopathies. Other variants cause no detectable pathology, and are thus considered non-pathological variants.

In the embryo:

- Gower 1 ($\xi_2\varepsilon_2$)
- Gower 2 ($\alpha_2\varepsilon_2$) (PDB: 1A9W)
- Hemoglobin Portland I ($\xi_2\gamma_2$)
- Hemoglobin Portland II ($\xi_2\beta_2$).

In the fetus:

- Hemoglobin F ($\alpha_2\gamma_2$) (PDB: 1FDH).

After birth:

- Hemoglobin A ($\alpha_2\beta_2$) (PDB: 1BZ0) – The most common with a normal amount over 95%
- Hemoglobin A_2 ($\alpha_2\delta_2$)—δ chain synthesis begins late in the third trimester and, in adults, it has a normal range of 1.5—3.5%
- Hemoglobin F ($\alpha_2\gamma_2$) – In adults Hemoglobin F is restricted to a limited population of red cells called F-cells. However, the level of Hb F can be elevated in persons with sickle-cell disease and beta-thalassemia.

Variant forms that cause disease:

- Hemoglobin D-Punjab — ($\alpha_2\beta^D{}_2$) – A variant form of hemoglobin.
- Hemoglobin H (β_4)— A variant form of hemoglobin, formed by a

tetramer of â chains, which may be present in variants of á thalassemia.

- Hemoglobin Barts (γ_4)—A variant form of hemoglobin, formed by a tetramer of ã chains, which may be present in variants of á thalassemia.
- Hemoglobin S ($\alpha_2\beta^S{}_2$) – A variant form of hemoglobin found in people with sickle cell disease. There is a variation in the â-chain gene, causing a change in the properties of hemoglobin, which results in sickling of red blood cells.
- Hemoglobin C ($\alpha_2\beta^C{}_2$)—Another variant due to a variation in the â-chain gene. This variant causes a mild chronic hemolytic anemia.
- Hemoglobin E ($\alpha_2\beta^E{}_2$)—Another variant due to a variation in the â-chain gene. This variant causes a mild chronic hemolytic anemia.
- Hemoglobin AS – A heterozygous form causing sickle cell trait with one adult gene and one sickle cell disease gene
- Hemoglobin SC disease – A compound heterozygous form with one sickle gene and another encoding Hemoglobin C.

DEGRADATION IN VERTEBRATE ANIMALS

When red cells reach the end of their life due to aging or defects, they are broken down in the spleen. The hemoglobin molecule is broken up, and the iron gets recycled. This process also produces one molecule of carbon monoxide for every molecule of heme degraded. Heme degradation is one of the few natural sources of carbon monoxide in the human body, and is responsible for the normal blood levels of carbon monoxide even in people breathing pure air. The other major final product of heme degradation is bilirubin. Increased levels of this chemical are detected in the blood if red cells are being destroyed more rapidly than usual. Improperly degraded hemoglobin protein or hemoglobin that has been released from the blood cells too rapidly can clog small blood vessels, especially the delicate blood filtering vessels of the kidneys, causing kidney damage. Iron is removed from heme and salvaged for later use, it is stored as hemosiderin or ferritin in tissues and transported in plasma by beta globulins as transferrins. When the porphyrin ring is broken up, the fragments are normally secreted as a yellow pigment called bilirubin, which is secreted into the intestines as bile. Intestines metabolise bilirubin into urobilinogen. Urobilinogen leaves the body in faeces, in a pigment called stercobilin. Globulin is metabolised into amino acids that are then released into circulation.

ROLE IN DISEASE

Hemoglobin deficiency can be caused either by a decreased amount of hemoglobin molecules, as in anemia, or by decreased ability of each molecule to bind oxygen at the same partial pressure of oxygen. Hemoglobinopathies (genetic defects resulting in abnormal structure of the hemoglobin molecule) may cause both. In any case, hemoglobin deficiency decreases blood oxygen-carrying capacity. Hemoglobin deficiency is, in general, strictly distinguished from hypoxemia, defined as decreased partial pressure of oxygen in blood, although both are causes of hypoxia (insufficient oxygen supply to tissues).

Other common causes of low hemoglobin include loss of blood, nutritional deficiency, bone marrow problems, chemotherapy, kidney failure, or abnormal hemoglobin (such as that of sickle-cell disease).

The ability of each hemoglobin molecule to carry oxygen is normally modified by altered blood pH or CO_2, causing an altered oxygen–hemoglobin dissociation curve. However, it can also be pathologically altered in, e.g., carbon monoxide poisoning.

Decrease of hemoglobin, with or without an absolute decrease of red blood cells, leads to symptoms of anemia. Anemia has many different causes, although iron deficiency and its resultant iron deficiency anemia are the most common causes in the Western world. As absence of iron decreases heme synthesis, red blood cells in iron deficiency anemia are hypochromic (lacking the red hemoglobin pigment) and microcytic (smaller than normal). Other anemias are rarer. In hemolysis (accelerated breakdown of red blood cells), associated jaundice is caused by the hemoglobin metabolite bilirubin, and the circulating hemoglobin can cause renal failure.

Some mutations in the globin chain are associated with the hemoglobinopathies, such as sickle-cell disease and thalassemia. Other mutations, as discussed at the beginning of the article, are benign and are referred to merely as hemoglobin variants. There is a group of genetic disorders, known as the porphyrias that are characterized by errors in metabolic pathways of heme synthesis. King George III of the United Kingdom was probably the most famous porphyria sufferer.

To a small extent, hemoglobin A slowly combines with glucose at the terminal valine (an alpha aminoacid) of each â chain. The resulting molecule is often referred to as Hb A_{1c}. As the concentration of glucose in the blood increases, the percentage of Hb A that turns into Hb A_{1c} increases. In diabetics whose glucose usually runs high, the percent Hb A_{1c} also runs high. Because of the slow rate of Hb A combination with glucose, the Hb A_{1c} percentage is representative of glucose level in the blood averaged over a longer time (the half-life of red blood cells, which is typically 50–55 days).

Glycosylated hemoglobin is the form of hemoglobin to which glucose is bound. The binding of glucose to amino acids in the hemoglobin takes place spontaneously (without the help of an enzyme) in many proteins, and is not known to serve a useful purpose. However, the binding to hemoglobin does serve as a record for average blood glucose levels over the lifetime of red cells, which is approximately 120 days. The levels of glycosylated hemoglobin are therefore measured in order to monitor the long-term control of the chronic disease of type 2 diabetes mellitus (T2DM). Poor control of T2DM results in high levels of glycosylated hemoglobin in the red blood cells. The normal reference range is approximately 4–5.9 %. Though difficult to obtain, values less than 7% are recommended for people with T2DM. Levels greater than 9% are associated with poor control of the glycosylated hemoglobin, and levels greater than 12% are associated with very poor control. Diabetics who keep their glycosylated hemoglobin levels close to 7% have a much better chance of avoiding the complications that may accompany diabetes (than those whose levels are 8% or higher). In addition, increased glycosylation of hemoglobin increases its affinity for oxygen, therefore preventing its release at the tissue and inducing a level of hypoxia in extreme cases.

Elevated levels of hemoglobin are associated with increased numbers or sizes of red blood cells, called polycythemia. This elevation may be caused by congenital heart disease, cor pulmonale, pulmonary fibrosis, too much erythropoietin, or polycythemia vera. High hemoglobin levels may also be caused by exposure to high altitudes, smoking, dehydration (artificially by concentrating Hb), advanced lung disease and certain tumors.

A recent study done in Pondicherry, India, shows its importance in coronary artery disease.

DIAGNOSTIC USES

Hemoglobin concentration measurement is among the most commonly performed blood tests, usually as part of a complete blood count. For example, it is typically tested before or after blood donation. Results are reported in g/L, g/dL or mol/L. 1 g/dL equals about 0.6206 mmol/L, although the latter units are not used as often due to uncertainty regarding the polymeric state of the molecule. This conversion factor, using the single globin unit molecular weight of 16,000 Da, is more common for hemoglobin concentration in blood. For MCHC (mean corpuscular hemoglobin concentration) the conversion factor 0.155, which uses the tetramer weight of 64,500 Da, is more common. Normal levels are:

- Men: 13.8 to 18.0 g/dL (138 to 180 g/L, or 8.56 to 11.17 mmol/L)
- Women: 12.1 to 15.1 g/dL (121 to 151 g/L, or 7.51 to 9.37 mmol/L)
- Children: 11 to 16 g/dL (111 to 160 g/L, or 6.83 to 9.93 mmol/L)
- Pregnant women: 11 to 14 g/dL 9.5 to 15(usual value during pregnancy)(110 to 140 g/L, or 6.83 to 8.69 mmol/L)

Normal values of hemoglobin in the 1st and 3rd trimesters of pregnant women must be at least 11 g/dL and at least 10.5 g/dL during the 2nd trimester.

Dehydration or hyperhydration can greatly influence measured hemoglobin levels. Albumin can indicate hydration status. If the concentration is below normal, this is called anemia. Anemias are classified by the size of red blood cells, the cells that contain hemoglobin in vertebrates. The anemia is called "microcytic" if red cells are small, "macrocytic" if they are large, and "normocytic" otherwise.

Hematocrit, the proportion of blood volume occupied by red blood cells, is typically about three times the hemoglobin concentration measured in g/dL. For example, if the hemoglobin is measured at 17 g/dL, that compares with a hematocrit of 51%.

Laboratory hemoglobin test methods require a blood sample (arterial, venous, or capillary) and analysis on hematology analyzer and CO-oximeter. Additionally, a new noninvasive hemoglobin (SpHb) test method called Pulse CO-Oximetry is also available with comparable accuracy to invasive methods. Concentrations of oxy- and deoxyhemoglobin can be measured continuously, regionally and noninvasively using NIRS. NIRS can be used both on the head as on muscles. This technique is often used for research in e.g. elite sports training, ergonomics, rehabilition, patient monitoring, neonatal research, functional brain monitoring, brain computer interface, urology (bladder contraction), neurology (Neurovascular coupling) and more. Long-term control of blood sugar concentration can be measured by the concentration of Hb A_{1c}. Measuring it directly would require many samples because blood sugar levels vary widely through the day. Hb A_{1c} is the product of the irreversible reaction of hemoglobin A with glucose. A higher glucose concentration results in more Hb A_{1c}. Because the reaction is slow, the Hb A_{1c} proportion represents glucose level in blood averaged over the half-life of red blood cells, is typically 50–55 days. An Hb A_{1c} proportion of 6.0% or less show good long-term glucose control, while values above 7.0% are elevated. This test is especially useful for diabetics.

The functional magnetic resonance imaging (fMRI) machine uses the signal from deoxyhemoglobin, which is sensitive to magnetic fields since it is paramagnetic. Combined measurement with NIRS shows good correlation with both the oxy- and deoxyhemoglobin signal compared to the bold signal.

HEMOGLOBINOMETER

A hemogoblinometer is a medical measuring device of hemoglobin blood concentration. It can operate by spectrophotometric measurement of hemoglobin concentration. Portable hemoglobinometers provide easy and convenient measurement of hematological variables, especially in areas where no clinic labs are available.

ANALOGUES IN NON-VERTEBRATE ORGANISMS

A variety of oxygen-transport and -binding proteins exist in organisms throughout the animal and plant kingdoms. Organisms including bacteria, protozoans, and fungi all have hemoglobin-like proteins whose known and predicted roles include the reversible binding of gaseous ligands. Since many of these proteins contain globins and the heme moiety (iron in a flat porphyrin support), they are often called hemoglobins, even if their overall tertiary structure is very different from that of vertebrate hemoglobin. In particular, the distinction of "myoglobin" and hemoglobin in lower animals is often impossible, because some of these organisms do not contain muscles. Or, they may have a recognizable separate circulatory system but not one that deals with oxygen transport (for example, many insects and other arthropods). In all these groups, heme/globin-containing molecules (even monomeric globin ones) that deal with gas-binding are referred to as oxyhemoglobins. In addition to dealing with transport and sensing of oxygen, they may also deal with NO, CO_2, sulfide compounds, and even O_2 scavenging in environments that must be anaerobic. They may even deal with detoxification of chlorinated materials in a way analogous to heme-containing P450 enzymes and peroxidases.

Fig. The Giant Tube Worm Riftia Pachyptila Showing Red Hemoglobin-Containing Plumes.

The structure of hemoglobins varies across species. Hemoglobin occurs in all kingdoms of organisms, but not in all organisms. Primitive species such as bacteria, protozoa, algae, and plants often have single-globin hemoglobins. Many nematode worms, molluscs, and crustaceans contain very large multisubunit molecules, much larger than those in vertebrates. In particular, chimeric hemoglobins found in fungi and giant annelids may contain both globin and other types of proteins.

One of the most striking occurrences and uses of hemoglobin in organisms is in the giant tube worm (Riftia pachyptila, also called Vestimentifera), which can reach 2.4 meters length and populates ocean volcanic vents. Instead of a digestive tract, these worms contain a population of bacteria constituting half the organism's weight. The bacteria react with H_2S from the vent and O_2 from the water to produce energy to make food from H_2O and CO_2. The worms end with a deep-red fan-like structure ("plume"), which extends into the water and absorbs H_2S and O_2 for the bacteria, and CO_2 for use as synthetic raw material similar to photosynthetic plants. The structures are bright-red due to their containing several extraordinarily complex hemoglobins that have up to 144 globin chains, each including associated heme structures. These hemoglobins are remarkable for being able to carry oxygen in the presence of sulfide, and even to carry sulfide, without being completely "poisoned" or inhibited by it as hemoglobins in most other species are.

OTHER OXYGEN-BINDING PROTEINS

MYOGLOBIN

Myoglobin is an iron- and oxygen-binding protein found in the muscle tissue of vertebrates in general and in almost all mammals. It is related to hemoglobin, which is the iron- and oxygen-binding protein in blood, specifically in the red blood cells. In humans, myoglobin is only found in the bloodstream after muscle injury. It is an abnormal finding, and can be diagnostically relevant when found in blood.

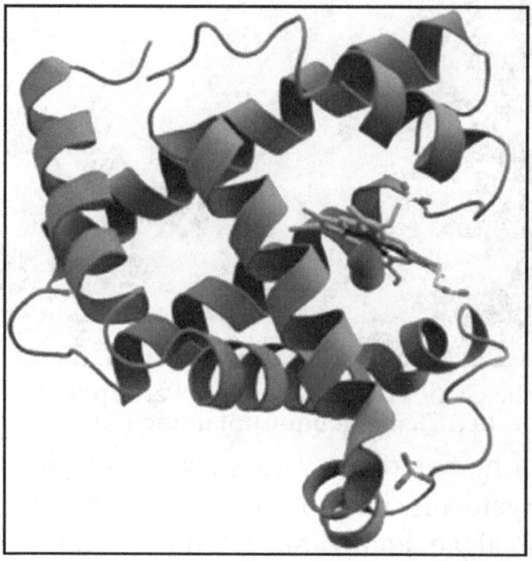

Fig. Model of Helical Domains in Myoglobin.

Myoglobin is the primary oxygen-carrying pigment of muscle tissues. High concentrations of myoglobin in muscle cells allow organisms to hold their breath for a longer period of time. Diving mammals such as whales and seals have muscles with particularly high abundance of myoglobin. Myoglobin is found in Type I muscle, Type II A and Type II B, but most texts consider myoglobin not to be found in smooth muscle.

Myoglobin was the first protein to have its three-dimensional structure revealed by X-ray crystallography. This achievement was reported in 1958 by John Kendrew and associates. For this discovery, John Kendrew shared the 1962 Nobel Prize in chemistry with Max Perutz. Despite being one of the most studied proteins in biology, its physiological function is not yet conclusively established: mice genetically engineered to lack myoglobin are viable, but showed a 30% reduction in volume of blood being pumped by the heart during a contraction. They adapted to this deficiency through natural reactions to inadequate oxygen supply (hypoxia) and a widening of blood vessels (vasodilation). In humans myoglobin is encoded by the MB gene.

Meat Color

Myoglobin contains hemes, pigments responsible for the color of red meat. The color that meat takes is partly determined by the degree of oxidation of the myoglobin. In fresh meat the iron atom is the ferrous state bound to a dioxygen molecule (O_2). Meat cooked well done is brown because the iron atom is now in the ferric (+3) oxidation state, having lost an electron. If meat has been exposed to nitrites, it will remain pink because the iron atom is bound to NO, nitric oxide (true of, e.g., corned beef or cured hams). Grilled meats can also take on a pink "smoke ring" that comes from the iron binding to a molecule of carbon monoxide.

Raw meat packed in a carbon monoxide atmosphere also shows this same pink "smoke ring" due to the same principles. Notably, the surface of this raw meat also displays the pink color, which is usually associated in consumers' minds with fresh meat. This artificially induced pink color can persist, reportedly up to one year. Hormel and Cargill are both reported to use this meat-packing process, and meat treated this way has been in the consumer market since 2003.

Role in Disease

Myoglobin is released from damaged muscle tissue (rhabdomyolysis), which has very high concentrations of myoglobin. The released myoglobin is filtered by the kidneys but is toxic to the renal tubular epithelium and so may cause acute renal failure. It is not the myoglobin itself that is toxic (it is a protoxin) but the ferrihemate portion that is dissociated from myoglobin in acidic environments (e.g., acidic urine, lysosomes).

Myoglobin is a sensitive marker for muscle injury, making it a potential marker for heart attack in patients with chest pain. However, elevated myoglobin has low specificity for acute myocardial infarction (AMI) and thus CK-MB, cTnT, ECG, and clinical signs should be taken into account to make the diagnosis.

Structure and Bonding

Myoglobin belongs to the globin superfamily of proteins, and as with other globins, consists of eight alpha helices connected by loops. Human globin contains 154 amino acids. Myoglobin contains a porphyrin ring with an iron at its center. A proximal histidine group (His-94) is attached directly to iron, and a distal histidine group (His-65) hovers near the opposite face. The distal imidazole is not bonded to the iron but is available to interact with the substrate O_2. This interaction encourages the binding of O_2, but not carbon monoxide (CO), which still binds about 240× more strongly than O_2.

The binding of O_2 causes substantial structural change at the Fe center, which shrinks in radius and moves into the center of N4 pocket. O_2-binding induces "spin-pairing": the five-coordinate ferrous deoxy form is high spin and the six coordinate oxy form is low spin and diamagnetic.

Synthetic Analogues

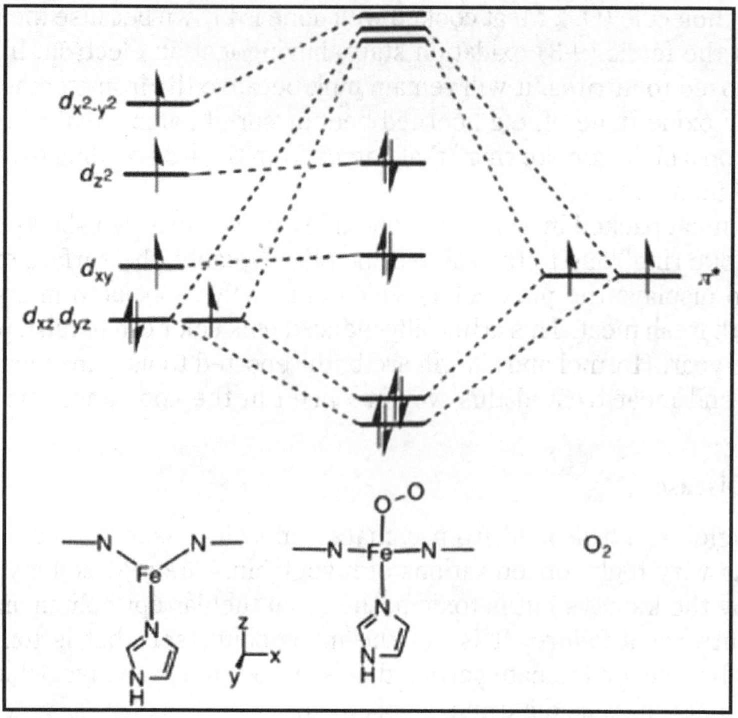

Fig. Molecular Orbital Description of Fe-O_2 Interaction in Myoglobin.

Many models of myoglobin have been synthesized as part of a broad interest in transition metal dioxygen complexes. A well known example is the picket fence porphyrin, which consists of a ferrous complex of a sterically bulky derivative of tetraphenylporphyrin. In the presence of an imidazole ligand, this ferrous complex reversibly binds O_2.

The O_2 substrate adopts a bent geometry, occupying the sixth position of the iron center. A key property of this model is the slow formation of the ì-oxo dimer, which is an inactive diferric state. In nature, such deactivation pathways are suppressed by protein matrix that prevents close approach of the Fe-porphyrin assemblies.

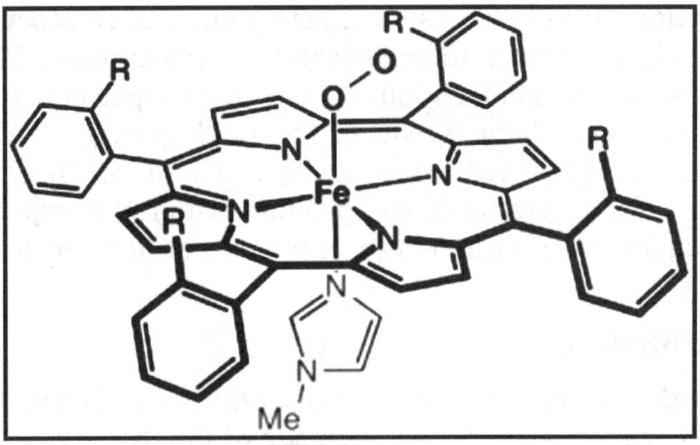

Fig. A Picket-Fence Porphyrin Complex of Fe, with Axial Coordination Sites Occupied by Methylimidazole (Green) and Dioxygen. The R Groups Flank the O_2-Binding Site.

HEMOCYANIN

Hemocyanins (also spelled haemocyanins) are proteins that transport oxygen throughout the bodies of some invertebrate animals. These metalloproteins contain two copper atoms that reversibly bind a single oxygen molecule (O_2). They are second only to hemoglobin in frequency of use as an oxygen transport molecule. Unlike the hemoglobin in red blood cells found in vertebrates, hemocyanins are not bound to blood cells but are instead suspended directly in the hemolymph. Oxygenation causes a color change between the colorless Cu(I) deoxygenated form and the blue Cu(II) oxygenated form.

Species Distribution

Hemocyanins are found only in the Mollusca and Arthropoda: the earliest hemocyanins were found in the snail Helix pomatia (a mollusc) and in the horseshoe crab (an arthropod). They were subsequently found to be common among crustaceans and are utilized by some land arthropods such as the tarantula Eurypelma californicum, the emperor scorpion, and the centipede Scutigera coleoptrata. Also, larval storage proteins in many insects appear to be derived from hemocyanins.

The Hemocyanin Superfamily

The arthropod hemocyanin superfamily is composed of phenoloxidases, hexamerins, pseudohemocyanins or cryptocyanins, (dipteran) hexamerin receptors. Phenoloxidase are copper containing tyrosinases. These proteins are involved in the process of sclerotization of arthropod cuticle, in wound healing, and humoral immune defense. Phenoloxidase is synthesized by zymogens and are activated by cleaving a N-terminal peptide.

Hexamerins are storage proteins commonly found in insects. These proteins are synthesized by the larval fat body and are associated with molting cycles or nutritional conditions Pseudohemocyanin and cryptocyanins genetic sequences are closely related to hemocyanins in crustaceans. These proteins have a similar structure and function, but lack the copper binding sites.

The evolutionary changes within the phylogeny of the hemocyanin superfamily are closely related to the emergence of these different proteins in various species. The understanding of proteins within this superfamily would not be well understood without the extensive studies of hemocyanin in arthropods.

Structure and Mechanism

Although the respiratory function of hemocyanin is similar to that of hemoglobin, there are a significant number of differences in its molecular structure and mechanism. Whereas hemoglobin carries its iron atoms in porphyrin rings (heme groups), the copper atoms of hemocyanin are bound as prosthetic groups coordinated by histidine residues. The active site of hemocyanin is composed of a pair of copper(I) cations which are directly coordinated to the protein through the driving force of imidazolic rings of six histidine residues. It has been noted that species using hemocyanin for oxygen transportation include crustaceans living in cold environments with low oxygen pressure. Under these circumstances hemoglobin oxygen transportation is less efficient than hemocyanin oxygen transportation. Nevertheless there are also terrestrial arthropods using hemocyanin, notably spiders and scorpions, that live in warm climates.

Most hemocyanins bind with oxygen non-cooperatively and are roughly one-fourth as efficient as hemoglobin at transporting oxygen per amount of blood. Hemoglobin binds oxygen cooperatively due to steric conformation changes in the protein complex, which increases hemoglobin's affinity for oxygen when partially oxygenated. In some hemocyanins of horseshoe crabs and some other species of arthropods, cooperative binding is observed, with Hill coefficients of 1.6 - 3.0. Hill coefficients vary depending on species and laboratory measurement settings. Hemoglobin, for comparison, has a Hill coefficient of usually 2.8 - 3.0. In these cases of cooperative binding hemocyanin was arranged in protein sub-complexes of 6 subunits (hexamer) each with one oxygen binding site; binding of oxygen on one unit in the complex would increase the affinity of the neighboring units. Each hexamer complex was arranged together to form a larger complex of dozens of hexamers. In one study, cooperative binding was found to be dependent on hexamers being arranged together in the larger complex, suggesting cooperative binding between hexamers. Hemocyanin oxygen-binding profile is also affected by dissolved salt ion levels and pH.

Hemocyanin is made of many individual subunit proteins, each of which contains two copper atoms and can bind one oxygen molecule (O_2). Each subunit weighs about 75 kilodaltons (kDa). Subunits may be arranged in dimers or hexamers depending on species; the dimer or hexamer complex is likewise arranged in chains or clusters with weights exceeding 1500 kDa. The subunits are usually homogeneous, or heterogeneous with two variant subunit types. Because of the large size of hemocyanin, it is usually found free-floating in the blood, unlike hemoglobin.

Hexamers are characteristic of arthropod hemocyanins. A hemocyanin of the tarantula Eurypelma californicum is made up of 4 hexamers or 24 pepide chains. A hemocyanin from the house centipede Scutigera coleoptrata is made up of 6 hexamers or 36 chains. Horseshoe crabs have an 8-hexamer (i. e. 48-chain) hemocyanin. Simple hexamers are found in the spiny lobster Panulirus interruptus and the isopod Bathynomus giganteus. Peptide chains in crustaceans are about 660 amino acid residues long, and in chelicerates they are about 625. In the large complexes there is a variety of variant chains, all about the same length; pure components do not usually self-assemble.

Catalytic Activity

Hemocyanin is homologous to the phenol oxidases (e.g. tyrosinase) since both enzymes share type 3 Cu active site coordination. In both cases inactive proenzymes such as hemocyanin, tyrosinase, and catcholoxidase must be activated first. This is done by removing the amino acid that blocks the entrance channel to the active site when the proenzyme is not active. There is currently no other known modifications necessary to activate the proenzyme and enable catalytic activity. Conformational differences determine the type of catalytic activity that the hemocyanin is able to perform. Hemocyanin also exhibits phenol oxidase activity, but with slowed kinetics from greater steric bulk at the active site. Partial denaturation actually improves hemocyanin's phenol oxidase activity by providing greater access to the active site.

Spectral Properties

Spectroscopy of oxyhemocyanin shows several salient features:

1. Resonance Raman spectroscopy shows symmetric binding
2. UV-Vis spectroscopy shows strong absorbances at 350 and 580 nm
3. OxyHc is EPR-silent indicating the absence of unpaired electrons
4. Infrared spectroscopy shows ν(O-O) of 755 cm"1

(1) rules out a mononuclear peroxo complex (2) does not match with the UV-Vis spectra of mononuclear peroxo and Kenneth Karlin's trans-peroxo models. (4) shows a considerably weaker O-O bond compared with Karlin's trans-peroxo model.

On the other hand, Nobumasa Kitajima's model shows í(O-O) of 741 cm"1 and UV-Vis absorbances at 349 and 551 nm, which agree with the experimental observations for oxyHc.

Anticancer Effects

The hemocyanin found in Concholepas concholepas blood has immunotherapeutic effects against bladder cancer in murine models. Researchers in 2006 primed mice with C. concholepas before implantation of bladder tumor (MBT-2) cells. Mice treated with C. concholepas hemocyanin showed antitumor effects: prolonged survival, decreased tumor growth and incidence, and lack of toxic effects and may have a potential use in future immunotherapy for superficial bladder cancer.

Keyhole limpet hemocyanin (KLH) is a known immune stimulant derived from circulating glycoproteins of the marine mollusk Megathura crenulata. KLH has been shown to be a significant treatment against the proliferations of breast cancer, pancreas cancer, and prostate cancer cells when delivered in vitro. The 2003 research study "Keyhole limpet hemocyanin, a novel immune stimulant with promising anticancer activity in Barrett's esophageal adenocarcinoma" shows that keyhole limpet hemocyanin is also effective in the inhibition of growth of human Barrett's esophageal cancer through both apoptic and nonapoptic mechanisms of cell death. This research is promising for future cancer curative research.

Case Studies: Environmental Impact on Hemocyanin Levels

A 2003 study of the effect of culture conditions of blood metabolites and hemocyanin of the white shrimp Litopenaeus vannamei found that the levels of hemocyanin, oxyhemocyanin in particular, are affected by the diet. The study compared oxyhemocyanin levels in the blood of white shrimp housed in an indoor pond with a commercial diet with that of white shrimp housed in an outdoor pond with a more readily available protein source (insects) as well. Oxyhemocyanin and blood glucose levels were higher in shrimp housed in outdoor ponds. It was also found that blood metabolite levels tended to be lower in low activity level species, such as crabs, lobsters, and the indoor shrimp when compared to the outdoor shrimp. This correlation is possibly indicative of the morphological and physiological evolution of crustaceans. The levels of these blood proteins and metabolites appear to be dependence on energetic demands and availability of those energy sources.

HEMERYTHRIN

Hemerythrin (also spelled haemerythrin; from Greek words αιμα= blood and ερυθροζ= red) is an oligomeric protein responsible for oxygen (O_2) transport in the marine invertebrate phyla of sipunculids, priapulids, brachiopods, and in a single annelid worm genus, Magelona. Myohemerythrin is a monomeric O_2-binding protein found in the muscles of marine invertebrates. Hemerythrin and myohemerythrin are essentially colorless when deoxygenated, but turn a violet-pink in the oxygenated state.

Hemerythrin does not, as the name might suggest, contain a heme. The names of the blood oxygen transporters hemoglobin, hemocyanin, hemerythrin, do not refer to the heme group (only found in globins), instead these names are derived from the Greek word for blood.

O_2 binding Mechanism

The mechanism of dioxygen binding is unusual. Most O_2 carriers operate via formation of dioxygen complexes, but hemerythrin holds the O_2 as a hydroperoxide. The site that binds O_2 consists of a pair of iron centres. The iron atoms are bound to the protein through the carboxylate side chains of a glutamate and aspartates as well as through five histidine residues. Hemerythrin and myohemerythrin are often described according to oxidation and ligation states of the iron centre:

Fe^{2+}—OH—Fe^{2+}	deoxy (reduced)
Fe^{2+}—OH—Fe^{3+}	semi-met
Fe^{3+}—O—Fe^{3+}—OOH"	oxy (oxidized)
Fe^{3+}—OH—Fe^{3+}— (any other ligand)	met (oxidized)

The uptake of O_2 by hemerythrin is accompanied by two-electron oxidation of the diferrous centre to produce a hydroperoxide (OOH") complex. The binding of O_2 is roughly described in this diagram:

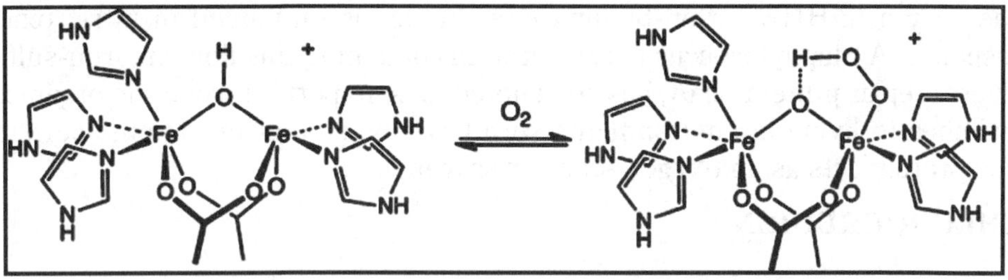

Fig... Active Site of Hemerythrin Before and After Oxygenation.

Deoxyhemerythrin contains two high-spin ferrous ions bridged by hydroxyl group (A). One iron is hexacoordinate and another is pentacoordinate. A hydroxyl group serves as a bridging ligand but also functions as a proton donor to the O_2 substrate. This proton-transfer result in the formation of a single oxygen atom (ì-oxo) bridge in oxy- and methemerythrin. O_2 binds to the pentacoordinate Fe^{2+} centre at the vacant coordination site (B). Then electrons are transferred from the ferrous ions to generate the binuclear ferric (Fe^{3+},Fe^{3+}) centre with bound peroxide (C).

Quaternary Structure and Cooperativity

Hemerythrin typically exists as a homooctamer or heterooctamer composed of á- and â-type subunits of 13-14 kDa each, although some species have dimeric, trimeric and tetrameric hemerythrins. Each subunit has a four-á-helix fold binding a binuclear iron centre. Because of its size hemerythrin is usually found in cells or "corpuscles" in the blood rather than free floating. Unlike hemoglobin, most hemerythrins lack cooperative binding to oxygen, making it roughly 1/4 as efficient as hemoglobin. In some brachiopods though, hemerythrin shows cooperative binding of O_2. Cooperative binding is achieved by interactions between subunits: the oxygenation of one subunit increases the affinity of a second unit for oxygen.

Hemerythrin affinity for carbon monoxide (CO) is actually lower than its affinity for O_2, unlike hemoglobin which has a very high affinity for CO. Hemerythrin's low affinity for CO poisoning reflects the role of hydrogen-bonding in the binding of O_2, a pathway mode that is incompatible with CO complexes which usually do not engage in hydrogen bonding.

Hemerythrin/HHE Cation-Binding Domain

The hemerythrin/HHE cation-binding domain occurs as a duplicated domain in hemerythrins, myohemerythrins and related proteins. This domain binds iron in hemerythrin, but can bind other metals in related proteins, such as cadmium in the Nereis diversicolor hemerythrin. It is also found in the NorA protein from Cupriavidus necator, this protein is a regulator of response to nitric oxide, which suggests a different set-up for its metal ligands. A protein from Cryptococcus neoformans (Filobasidiella neoformans) that contains haemerythrin/HHE cation-binding domains is also involved in nitric oxide response. A Staphylococcus aureus protein containing this domain, iron-sulfur cluster repair protein ScdA, has been noted to be important when the organism switches to living in environments with low oxygen concentrations; perhaps this protein acts as an oxygen store or scavenger.

CHLOROCRUORIN

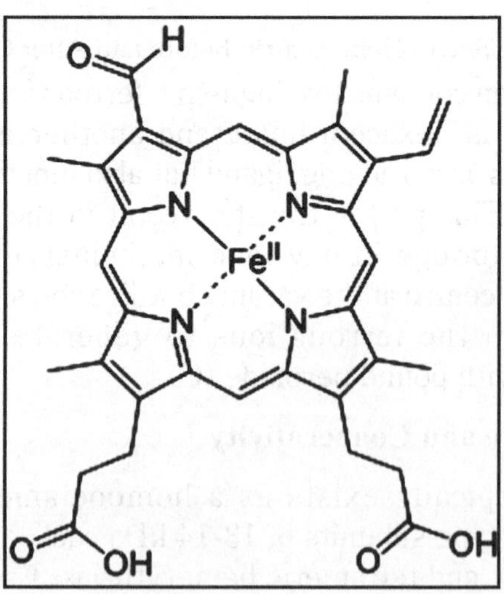

Fig. Chlorocruorin.

Chlorocruorin is an oxygen-binding hemeprotein present in the blood plasma of many annelids, particularly certain marine polychaetes. Its affinity for oxygen is weaker than that of most haemoglobins. A dichromatic compound, chlorocruorin is noted for appearing green in dilute solutions, though it appears light red when found in concentrated solutions.

Its structure is very similar to erythrocruorin (which is likewise very similar to multiple subunits of myoglobin) and it contains many 16-17kDa myoglobin-like subunits arranged in a giant complex of over a hundred subunits with interlinking proteins as well with a total weight exceeding 3600kDa. The only significant difference between chlorocruorin and erythrocruorin is that chlorocruorin carries an abnormal heme group structure. This enormous macromolecule is typically found free floating in the plasma, and not contained within red blood cells.

VANABINS

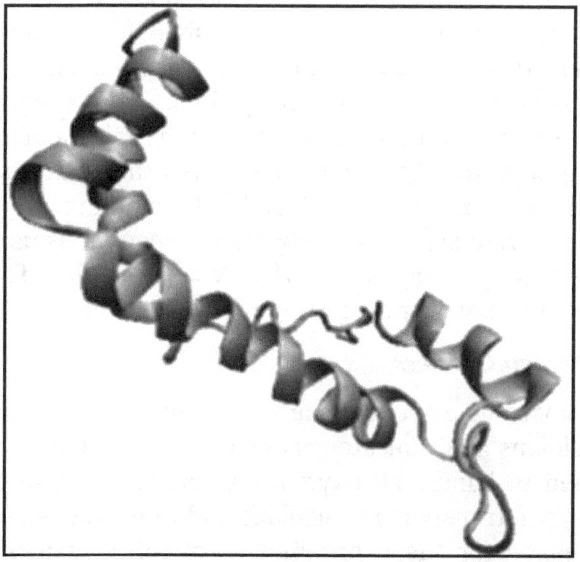

Fig. 3-D Representation of the Structure of Vanabin2 from Ascisia Sydneiensis Samea.

Vanabins (also known as vanadium-associated proteins or vanadium chromagen) are a specific group of vanadium-binding metalloproteins. Vanabins are found almost exclusively in the blood cells, or vanadocytes of some ascidians and tunicates (sea squirts). The vanabins extracted from tunicate vanadocytes are often called hemovanadins. These organisms are able to concentrate vanadium to a level more than 100 times higher than in the surrounding seawater. Vanabin proteins seem to be involved in collecting and accumulating this metal ion. At present there is no conclusive understanding of why these organisms collect vanadium, and it remains a biological mystery. German chemist Martin Henze discovered vanadium in ascidiaceans in 1911.

Theories

Vanabins as Oxygen Carriers

Because of the high concentration of vanadium in the blood, it has been assumed that vanabins are used for oxygen transport like iron-based hemoglobin or copper-based hemocyanin. Unfortunately no scientific evidence can be found to back this hypothesis.

The highest concentration of vanadium found so far, 350 mM, was found in the blood cells of Ascidia gemmata belonging to the suborder Phlebobranchia. This concentration is 10,000,000 times higher than that in seawater.

Vanabins accumulate vanadium in the blood cells and produce V(III) species and vandanyl ions (V(IV)) from orthovanadate ions (V(V)), with the use of NADPH as a reducing agent. Vanabins also transport the reduced vanadium species into the vacuoles of vanadocytes (vanadium-containing blood cells). The vacuoles are kept at a very acidic pH of 1.9, made possible by pumping hydrogen ions into the vacuoles by use of energy intensive H^+-ATPase. All the use of NADPH and ATP to collect and maintain the vanadium is extremely energy intensive, unusual for oxygen transporting proteins. Other oxygen transporting proteins have a very low dissociation constant with their metal prosthetic group and bind these groups tightly. Vanabins on the other hand have a moderate dissociation constant and do not tightly bind vanadium. Most importantly, because of this moderate dissociation constant, vanadium is usually found free-floating and separated from any proteins inside the vacuoles. This is completely different from other oxygen transporting proteins.

Vanabins Not Oxygen Carriers

The most convincing evidence against vanadium in use for oxygen transport is that many ascidians and tunicates also have hemocyanin in their blood that could be assumed to handle all oxygen transport. The use of vanabins and vanadium for oxygen transport in ascidians and tunicates is doubtful. Another hypothesized reason for these organisms collecting vanadium is to make themselves toxic to predators, parasites and microorganisms.

Technological use of Vanabins

The world oceans carry diluted concentrations of rare minerals. Total amount of gold, silver, manganese and platinum (among others) in all the world's oceans rivals that of terrestrial mines, yet because of the low concentration of these minerals in sea water, extraction has been considered impractical. Understanding the mechanisms behind how some organisms concentrate rare minerals (such as siderophores) from sea water could make it biotechnologically possible to grow and farm organisms to filter sea water and extract these minerals, similar to bioleaching.

ERYTHROCRUORIN

Erythrocruorin is a large oxygen-carrying protein, whose molecular mass is greater than 3.5 million Daltons. It is related to the similar chlorocruorin. It is found in many annelids and arthropods (including some insects).

LEGHEMOGLOBIN

Leghemoglobin (also leghaemoglobin or legoglobin) is a nitrogen or oxygen carrier, because naturally occurring oxygen and nitrogen interact similarly with this protein; and a hemoprotein found in the nitrogen-fixing root nodules of leguminous plants. It is produced by legumes in response to the roots being colonized by nitrogen-fixing bacteria, termed rhizobia, as part of the symbiotic interaction between plant and bacterium: roots not colonized by Rhizobium do not synthesise leghemoglobin. Leghemoglobin has close chemical and structural similarities to hemoglobin, and, like hemoglobin, is red in colour. The holoprotein (protein + heme cofactor) is widely believed to be a product of both plant and the bacterium in which the apoprotein is produced by the plant and the heme (an iron atom bound in a porphyrin ring) is produced by the bacterium. There is some evidence, however, that suggests that the heme moiety is also produced by the plant.

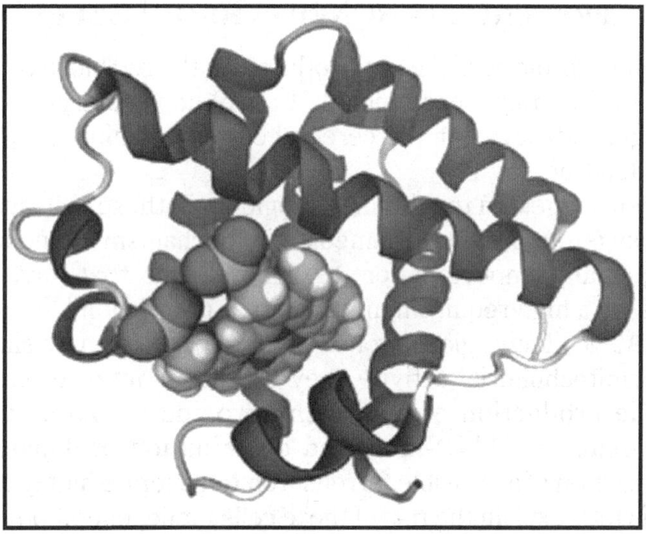

Fig. Leghemoglobin from Soybean.

In plants infected with Rhizobium, (such as alfalfa or soybeans), the presence of oxygen in the root nodules would reduce the activity of the oxygen-sensitive nitrogenase - an enzyme responsible for the fixation of atmospheric nitrogen. Leghemoglobin buffers the concentration of free oxygen in the cytoplasm of infected plant cells to ensure the proper function of root nodules. Leghemoglobin has a high affinity for oxygen (a K_m of about 0.01 μM), about ten times higher than the â chain of human hemoglobin. This allows an oxygen concentration that is low enough to allow nitrogenase to function but high enough that it can provide the bacteria with oxygen for respiration.

Although leghemoglobin was once thought to provide a buffer for nodule oxygen, recent studies indicate that it stores only enough oxygen to support nodule respiration for a few seconds. Its function is to help provide oxygen to the respiring symbiotic bacterial cells in a manner analogous to hemoglobin transporting oxygen to respiring tissues in animals. Other plants, like Casuarina sp which is an actinorhizal plant, produce a hemoglobin in their symbiotic root nodules.

COBOGLOBIN

A coboglobin is a synthetic compound, a metalloprotein chemically similar to hemoglobin or myoglobin but using the metal cobalt instead of iron (hence the name). Just like hemoglobin and myoglobin, the coboglobins are able to reversibly bind molecular oxygen (O_2) at the metal atom. However they lose this ability much faster than the natural molecules. Blood of this type would be amber yellow in colour when in the veins while uncoloured and clear in the arteries.

PRESENCE IN NONERYTHROID CELLS

Some nonerythroid cells (i.e., cells other than the red blood cell line) contain hemoglobin. In the brain, these include the A9 dopaminergic neurons in the substantia nigra, astrocytes in the cerebral cortex and hippocampus, and in all mature oligodendrocytes.

It has been suggested that brain hemoglobin in these cells may enable the "storage of oxygen to provide a homeostatic mechanism in anoxic conditions, which is especially important for A9 DA neurons that have an elevated metabolism with a high requirement for energy production". It has been noted further that "A9 dopaminergic neurons may be at particular risk since in addition to their high mitochondrial activity they are under intense oxidative stress caused by the production of hydrogen peroxide via autoxidation and/or monoamine oxidase (MAO)-mediated deamination of dopamine and the subsequent reaction of accessible ferrous iron to generate highly toxic hydroxyl radicals". This may explain the risk of these cells for degeneration in Parkinson's disease.

The hemoglobin-derived iron in these cells is not the cause of the post-mortem darkness of these cells (origin of the Latin name, substantia nigra), but rather is due to neuromelanin.

Outside the brain, hemoglobin has non-oxygen-carrying functions as an antioxidant and a regulator of iron metabolism in macrophages, alveolar cells, and mesangial cells in the kidney.

HEMOGLOBIN VARIANTS

GLYCATED HEMOGLOBIN

Glycated hemoglobin (hemoglobin A1c, HbA_{1c}, A1C, or Hb_{1c}; sometimes also HbA1c or HGBA1C) is a form of hemoglobin that is measured primarily to identify the average plasma glucose concentration over prolonged periods of time. It is formed in a non-enzymatic glycation pathway by hemoglobin's exposure to plasma glucose. HbA_{1c} is a measure of the beta-N-1-deoxy fructosyl component of hemoglobin. Normal levels of glucose produce a normal amount of glycated hemoglobin. As the average amount of plasma glucose increases, the fraction of glycated hemoglobin increases in a predictable way. This serves as a marker for average blood glucose levels over the previous 3 months prior to the measurement as this is the half life of red blood cells. In diabetes mellitus, higher amounts of glycated hemoglobin, indicating poorer control of blood glucose levels, have been associated with cardiovascular disease, nephropathy, and retinopathy. Monitoring HbA_{1c} in type 1 diabetic patients may improve outcomes.

Hemoglobin A1c was first separated from other forms of hemoglobin by Huisman and Meyering in 1958 using a chromatographic column. It was first characterized as a glycoprotein by Bookchin and Gallop in 1968. Its increase in diabetes was first described in 1969 by Samuel Rahbar et al. The reactions leading to its formation were characterized by Bunn and his coworkers in 1975.

The use of hemoglobin A1c for monitoring the degree of control of glucose metabolism in diabetic patients was proposed in 1976 by Anthony Cerami, Ronald Koenig and coworkers.

Principle

Glycation of proteins is a frequent occurrence, but in the case of hemoglobin, a nonenzymatic reaction occurs between glucose and the N-end of the beta chain. This forms a Schiff base which is itself converted to 1-deoxyfructose. This rearrangement is an example of an Amadori rearrangement. When blood glucose levels are high, glucose molecules attach to the hemoglobin in red blood cells. The longer hyperglycemia occurs in blood, the more glucose binds to hemoglobin in the red blood cells and the higher the glycated hemoglobin.

Once a hemoglobin molecule is glycated, it remains that way. A buildup of glycated hemoglobin within the red cell, therefore, reflects the average level of glucose to which the cell has been exposed during its life-cycle. Measuring glycated hemoglobin assesses the effectiveness of therapy by monitoring long-term serum glucose regulation.

The HbA_{1c} level is proportional to average blood glucose concentration over the previous four weeks to three months. Some researchers state that the major proportion of its value is weighted toward the most recent 2 to 4 weeks. This is also supported by data from practice showing that HbA1c levels improved significantly by 20 days from the start of the intensification of glucose-lowering treatment.

Measuring HbA_{1c}

A number of techniques are used to measure hemoglobin A1c.
Laboratories use:

- High-performance liquid chromatography (HPLC): The HbA_{1c} result is calculated as a ratio to total hemoglobin by using a chromatogram.
- Immunoassay
- Enzymatic
- Capillary electrophoresis
- Boronate affinity chromatography

Point of care (e.g., doctor's office) devices use:

- Immunoassay
- Boronate affinity chromatography

In the United States, HbA_{1c} testing laboratories are certified by the National Glycohemoglobin Standardization Program (NGSP) to standardise them against the results of the 1993 Diabetes Control and Complications Trial (DCCT). An additional percentage scale, Mono S is in use by Sweden and KO500 is in Japan.

Switch to IFCC Units

The American Diabetes Association (ADA), European Association for the Study of Diabetes (EASD) and International Diabetes Federation (IDF) have agreed that, in the future, HbA_{1c} is to be reported in the International Federation of Clinical Chemistry (IFCC) units. IFCC reporting was introduced in Europe except for the UK in 2003; the UK carried out dual reporting from 1 June 2009 until 1 October 2011.

Conversion between DCCT and IFCC is by the following equation:

$$IFCC\ HBA1c\,(mmol/mol) = [DCCT\ HBA1c\,(\%) - 2.15] * 10.929$$

IFCC-HbA_{1c}	DCCT-HbA_{1c}	Mono S- HbA_{1c}

Interpretation of Results

Laboratory results may differ depending on the analytical technique, the age of the subject, and biological variation among individuals. Two individuals with the same average blood sugar can have HbA_{1c} values that differ by as much as 3 percentage points.

Results can be unreliable in many circumstances, such as after blood loss, for example, after surgery, blood transfusions, anemia, or high erythrocyte turnover; in the presence of chronic renal or liver disease; after administration of high-dose vitamin C; or erythropoetin treatment. In general, the reference range (that found in healthy persons), is about 20–40 mmol/mol (4–5.9 DCCT %).

Higher levels of HbA_{1c} are found in people with persistently elevated blood sugar, as in diabetes mellitus. While diabetic patient treatment goals vary, many include a target range of HbA_{1c} values. A diabetic person with good glucose control has a HbA_{1c} level that is close to or within the reference range.

The International Diabetes Federation and the American College of Endocrinology recommend HbA_{1c} values below 48 mmol/mol (6.5 DCCT %), while the American Diabetes Association recommends HbA_{1c} be below 53 mmol/mol (7.0 DCCT %) for most patients. Recent results from large trials suggest that a target below 53 mmol/mol (7 DCCT %) for older adults with type 2 diabetes may be excessive: Below 53 mmol/mol (7 DCCT %) the health benefits of reduced A1C become smaller, and the intensive glycemic control required to reach this level leads to an increased rate of dangerous hypoglycemic episodes.

A retrospective study of 47,970 old type 2 diabetes patients found that patients with an HbA_{1c} more than 48 mmol/mol (6.5 DCCT %) had an increased mortality rate, but a later international study contradicted these findings.

A review of the UKPDS accord, advance and VADT trials estimated that the risks of the main complications of diabetes (retinopathy, nephropathy, neuropathy and macrovascular disease) decreased by approximately 3% for every 1 mmol/mol decrease in HbA_{1c}.

Practitioners must consider an individual patient's health, their risk of hypoglycemia, and their specific health risks when setting a target HbA_{1c} level. Because patients are responsible for averting or responding to their own hypoglycemic episodes, the patient's input and the doctor's assessment of the patient's self-care skills are also important.

Persistent elevations in blood sugar (and, therefore, HbA_{1c}) increase the risk of long-term vascular complications of diabetes such as coronary disease, heart attack, stroke, heart failure, kidney failure, blindness, erectile dysfunction, neuropathy (loss of sensation, especially in the feet), gangrene, and gastroparesis (slowed emptying of the stomach). Poor blood glucose control also increases the risk of short-term complications of surgery such as poor wound healing.

Lower-than-expected levels of HbA_{1c} can be seen in people with shortened red blood cell lifespan, such as with glucose-6-phosphate dehydrogenase deficiency, sickle-cell disease, or any other condition causing premature red blood cell death. Blood donation will result in rapid replacement of lost RBCs with newly formed red blood cells. Since these new RBCs will have only existed for a short period of time, their presence will lead HbA_{1c} to underestimate the actual average levels. There may also be distortions resulting from blood donation which occurred as long as two months before due to an abnormal synchronization of the age of the RBCs, resulting in an older than normal average blood cell life (resulting in an overestimate of actual average blood glucose levels). Conversely, higher-than-expected levels can be seen in people with a longer red blood cell lifespan, such as with Vitamin B_{12} or folate deficiency.

The approximate mapping between HbA_{1c} values given in DCCT percentage (%) and eAG (estimated average glucose) measurements is given by the following equation:

eAG(mg/dl) = 28.7 × A1C " 46.7

eAG(mmol/l) = 1.59 × A1C " 2.59

Data in parentheses are 95% confidence intervals

HbA_{1c}		**eAG (estimated average glucose)**	
(%)	**(mmol/mol)**	**(mmol/L)**	**(mg/dL)**
5	31	5.4 (4.2–6.7)	97 (76–120)
6	42	7.0 (5.5–8.5)	126 (100–152)
7	53	8.6 (6.8–10.3)	154 (123–185)
8	64	10.2 (8.1–12.1)	183 (147–217)
9	75	11.8 (9.4–13.9)	212 (170–249)
10	86	13.4 (10.7–15.7)	240 (193–282)
11	97	14.9 (12.0–17.5)	269 (217–314)
12	108	16.5 (13.3–19.3)	298 (240–347)
13	119	18.1 (15–21)	326 (260–380)
14	130	19.7 (16–23)	355 (290–410)
15	140	21.3 (17–25)	384 (310–440)
16	151	22.9 (19–26)	413 (330–480)
17	162	24.5 (20–28)	441 (460–510)
18	173	26.1 (21–30)	470 (380–540)
19	184	27.7 (23–32)	499 (410–570)

The 2010 American Diabetes Association Standards of Medical Care in Diabetes added the HbA_{1c} e" 48 mmol/mol (e"6.5 DCCT %) as another criterion for the diagnosis of diabetes.

Indications and Use

Glycated hemoglobin testing is recommended for both (a) checking the blood sugar control in people who might be pre-diabetic and (b) monitoring blood sugar control in patients with more elevated levels, termed diabetes mellitus. There is a significant proportion of people who are unaware of their elevated HbA_{1c} level before they have blood lab work. For a single blood sample, it provides far more revealing information on glycemic behavior than a fasting blood sugar value. However, fasting blood sugar tests are crucial in making treatment decisions. The American Diabetes Association guidelines are similar to others in advising that the glycated hemoglobin test be performed at least two times a year in patients with diabetes that are meeting treatment goals (and that have stable glycemic control) and quarterly in patients with diabetes whose therapy has changed or that are not meeting glycemic goals.

Glycated hemoglobin measurement is not appropriate where there has been a change in diet or treatment within 6 weeks. Likewise, the test assumes a normal red blood cell aging process and mix of hemoglobin subtypes (predominantly HbA in normal adults). Hence, people with recent blood loss, hemolytic anemia, or genetic differences in the hemoglobin molecule (hemoglobinopathy) such as sickle-cell disease and other conditions, as well as those that have donated blood recently, are not suitable for this test.

Due to glycated hemoglobin's variability (as shown in the table above), additional measures should be checked in patients at or near recommended goals. People with HbA_{1c} values at 64 mmol/mol or less should be provided additional testing to determine whether the HbA_{1c} values are due to averaging out high blood glucose (hyperglycemia) with low blood glucose (hypoglycemia) or the HbA_{1c} is more reflective of an elevated blood glucose that does not vary much throughout the day.

Devices such as continuous blood glucose monitoring allow people with diabetes to determine their blood glucose levels on a continuous basis, testing every few minutes. Continuous use of blood glucose monitors is becoming more common, and the devices are covered by many health insurance plans but not by Medicare in the United States. The supplies tend to be expensive, since the sensors must be changed at least weekly. Another test that is useful in determining if HbA_{1c} values are due to wide variations of blood glucose throughout the day is 1,5 Anhydroglucitol, also known as GlycoMark. GlycoMark reflects only the times that the person experiences hyperglycemia above 180 mg/dL over a two-week period.

Concentrations of hemoglobin A1 (HbA1) are increased, both in diabetic patients and in patients with renal failure, when measured by ion-exchange chromatography. The thiobarbituric acid method (a chemical method specific for the detection of glycation) shows that patients with renal failure have values for glycated hemoglobin similar to those observed in normal subjects, suggesting that the high values in these patients are a result of binding of something other than glucose to hemoglobin.

In autoimmune hemolytic anemia, concentrations of hemoglobin A1 (HbA1) is undetectable. Administration of prednisolone (PSL) will allow the HbA1 to be detected. The alternative fructosamine test may be used in these circumstances and it also reflects an average of blood glucose levels over the preceding 2 to 3 weeks.

All the major institutions like International Expert Committee Report, drawn from the International Diabetes Federation (IDF), the European Association for the Study of diabetes (EASD), and the American Diabetes Association (ADA), suggests the HbA_{1c} level of 48 mmol/mol (6.5 DCCT %) as a diagnostic level. The Committee Report further states that, when HbA_{1c} testing cannot be done, the fasting and glucose tolerance tests be done.

Diagnosis of diabetes during pregnancy continues to require fasting and glucose tolerance measurements for gestational diabetes, and not the glycated hemoglobin.

Modification by Exercise Training

A meta-analysis of research done to identify the effect of two different kinds of training programs (combined aerobic and eccentric resistance exercise program and aerobic exercise only) on the glycated hemoglobin levels of individuals with T2DM found that the effect of combining resistance exercise with aerobic exercise improved the glucose control more than just the aerobics alone, but it should be noted that BMI of the resistance plus aerobic exercise group decreased more than the aerobic only group.

The average effect of the training programs included reductions of glycated hemoglobin of 9 mmol/mol (0.8 percentage points), which was a result similar to that of long-term diet and drug or insulin therapy (which result in a reduction of 6.5–9.0 mmol/mol (i.e. 0.6–0.8 points).

Standardization and Traceability

Hemoglobin A1c is now standardized & traceable to IFCC methods HPLC-CE & HPLC-MS. A new unit (mmol/mol) is used as part of this standardization. The standardized test does not however test for Iodine levels in the blood. Hypothyroidism or Iodine supplementation are known sources that artificially raise the A1c number.

HEMOGLOBIN A2

Hemoglobin A2 (HbA_2) is a normal variant of hemoglobin A that consists of two alpha and two delta chains ($á_2ä_2$) and is found at low levels in normal human blood. Hemoglobin A2 may be increased in beta thalassemia or in people who are heterozygous for the beta thalassemia gene. HbA2 exists in small amounts in all adult humans (1.5-3.1% of all hemoglobin molecules) and is increased in people with Sickle-cell disease. Its biological importance is not yet known.

HEMOGLOBIN C

Hemoglobin C (abbreviated as Hb C or HbC) is an abnormal hemoglobin in which substitution of a glutamic acid residue with a lysine residue at the 6th position of the â-globin chain has occurred (E6K substitution).

Clinical Significance

This mutated form reduces the normal plasticity of host erythrocytes causing a hemoglobinopathy. In those who are heterozygous for the mutation, about 28–44% of total hemoglobin (Hb) is HbC, and no anemia develops. In homozygotes, nearly all Hb is in the HbC form, resulting in mild hemolytic anemia

Presentation

Target cells, microspherocytes and HbC crystals are found in a blood smear from a homozygous patient.

Combinations with Other Conditions

Individuals with sickle cell–hemoglobin C (HbSC), have the gene for HbS inherited from one parent and the gene for HbC is inherited from the other parent: they are "heterozygous". Since HbC does not polymerize as readily as HbS, there is less sickling (fewer sickle cells). The peripheral smear demonstrates mostly target cells and only a few sickle cells. There are fewer acute vaso-occlusive events. However, persons with hemoglobin SC disease (HbSC) have more significant retinopathy, ischemic necrosis of bone, and priapism than those with pure SS disease. People with hemoglobin C trait have red blood cells that have normal hemoglobin A and an abnormal hemoglobin. The abnormal hemoglobin is called hemoglobin C. People with hemoglobin C trait have slightly more hemoglobin A than hemoglobin C. People with Hemoglobin C trait do not have health problems related to having the trait. People with hemoglobin C do not have Hemoglobin C disease or sickle cell disease. They cannot develop these diseases later in life. They can pass hemoglobin C trait to their offspring. Individuals who carry the hemoglobin C trait can have a child with Hemoglobin C disease or Hemoglobin SC disease. Hemoglobin C disease is not a form of sickle cell disease.

Symptoms

Most people do not have symptoms. It can cause a mild to moderate enlargement of the spleen, splenomegaly, as well as hemolytic anemia (which is the form of anemia due to abnormal breakdown of red blood cells prematurely). Too much hemoglobin C can reduce the number and size of red blood cells in your body, causing mild anemia. Occasionally, jaundice may occur. Some persons with this disease may develop gallstones that require treatment. Continued hemolysis may produce pigmented gallstones, an unusual type of gallstone composed of the dark-colored contents of red blood cells.

Epidemiology

Hemoglobin C gene is found in 2-3% of US Blacks while 8% of US Blacks have hemoglobin S (Sickle) gene. Thus Hemoglobin SC disease is significantly more common than Hemoglobin CC disease. Hemoglobin C is found in areas of West Africa, such as Ghana, where Yorubas live. About 1 out of every 40 African-Americans has hemoglobin C trait. The trait also affects people whose ancestors came from Italy, Greece, Africa, Latin America, and the Caribbean region. However, it is possible for a person of any race or nationality to have hemoglobin C trait. In terms of geographic distribution, the hemoglobin C allele is found at the highest frequencies in West Africa, where it has been associated with protection against malaria. Hemoglobin C disease is present at birth, though some cases may not be diagnosed until adulthood. Both sexes, male and female, are affected equally.

Heredity of Hemoglobin C Disease

Hemoglobin C disease is an autosomal recessive disorder that results from the biparental inheritance of the allele that encodes for hemoglobin C. If both parents are carriers of hemoglobin C, there is a chance of having a child with hemoglobin C disease. These are the possible outcomes with each pregnancy. 25 percent (1 in 4) chance of having a child with hemoglobin C disease 50 percent (1 in 2) chance of having a child who is a carrier of hemoglobin C 25 percent (1 in 4) chance of having a child without disease and is not a carrier.

Clinical Tests

Physical Examination may show an enlarged spleen. Tests that may be done include: Complete Blood Count (CBC), Hemoglobin electrophoresis, Peripheral blood smear, and Blood hemoglobin.

Treatment

Usually no treatment is needed. Folic acid supplementation may help produce normal red blood cells and improve the symptoms of anemia

Prognosis

Overall, hemoglobin C disease is one of the more benign hemoglobinopathies. Mild-to-moderate reduction in RBC lifespan may accompany from mild hemolytic anemia. Individuals with hemoglobin C disease have sporadic episodes of musculoskeletal (joint) pain. People with hemoglobin C disease can expect to lead a normal life.

Prevention

Genetic counseling may be appropriate for high-risk couples who wish to have a baby.

FETAL HEMOGLOBIN

Fetal hemoglobin, or foetal haemoglobin, (also hemoglobin F, HbF, or $\acute{a}_2\tilde{a}_2$) is the main oxygen transport protein in the human fetus during the last seven months of development in the uterus and persists in the newborn until roughly 6 months old. Functionally, fetal hemoglobin differs most from adult hemoglobin in that it is able to bind oxygen with greater affinity than the adult form, giving the developing fetus better access to oxygen from the mother's bloodstream. In newborns, fetal hemoglobin is nearly completely replaced by adult hemoglobin by approximately 6 months postnatally, except in a few thalassemia cases in which there may be a delay in cessation of HbF production until 3–5 years of age. In adults, fetal hemoglobin production can be reactivated pharmacologically, which is useful in the treatment of diseases such as sickle-cell disease.

Oxygenated blood is delivered to the fetus via the umbilical vein from the placenta, which is anchored to the wall of the mother's uterus. The chorion acts as a barrier between the maternal and fetal circulation so that there is no admixture of maternal and fetal blood. Blood in the maternal circulation is delivered via open ended arterioles to the intervillous space of the chorionic plate, where it bathes the chorionic villi that carry umbilical capillary beds, thereby allowing gas exchange to occur between the maternal and fetal circulation. Deoxygenated maternal blood drains into open ended intervillous venules to return to maternal circulation. Due to the admixture of oxygenated and deoxygenated blood, maternal blood in the intervillous space is lower in oxygen than arterial blood. As such, fetal hemoglobin must be able to bind oxygen with greater affinity than adult hemoglobin in order to compensate for the relatively lower oxygen tension of the maternal blood supplying the chorion.

Fetal hemoglobin's affinity for oxygen is substantially greater than that of adult hemoglobin. Notably, the P50 value for fetal hemoglobin is lower than adult hemoglobin (i.e., the partial pressure of oxygen at which the protein is 50% saturated; lower values indicate greater affinity). The P50 of fetal hemoglobin is roughly 19 mmHg, whereas adult hemoglobin is approximately 26.8 mmHg. As a result, the "oxygen saturation curve", which plots percent saturation vs. pO_2, is left-shifted for fetal hemoglobin as compared to adult hemoglobin.

This greater affinity for oxygen is explained by the lack of fetal hemoglobin's interaction with 2,3-bisphosphoglycerate (2,3-BPG or 2,3-DPG). In adult red blood cells, this substance decreases the affinity of hemoglobin for oxygen. 2,3-BPG is also present in fetal red blood cells, but interacts less efficiently with fetal hemoglobin than adult hemoglobin. This is due to a change in a single amino acid (residue 143) found in the 2,3-BPG 'binding pocket': from histidine to serine, which gives rise to the greater oxygen affinity.

Whereas histidine is positively charged and interacts well with the negative charges found on the surface of 2,3-BPG, Serine has a neutrally charged side chain at physiological pH, and interacts less well. This change results in less binding of 2,3-BPG to fetal Hb, and as a result oxygen will bind to it with higher affinity than adult hemoglobin.

For mothers to deliver oxygen to a fetus, it is necessary for the fetal hemoglobin to extract oxygen from the maternal oxygenated hemoglobin across the placenta. The higher oxygen affinity required for fetal hemoglobin is achieved by the protein subunit ã (gamma), instead of the â (beta) subunit. Because the ã subunit has fewer positive charges than the (adult) â subunit, 2,3-BPG is less electrostatically bound to fetal hemoglobin compared to adult hemoglobin. This lowered affinity allows for adult hemoglobin (maternal hemoglobin) to readily transfer its oxygen to the fetal bloodstream.

Distribution

After the first 10 to 12 weeks of development, the fetus' primary form of hemoglobin switches from embryonic hemoglobin to fetal hemoglobin. At birth, fetal hemoglobin comprises 50-95% of the infant's hemoglobin. These levels decline after six months as adult hemoglobin synthesis is activated while fetal hemoglobin synthesis is deactivated. Soon after, adult hemoglobin (hemoglobin A in particular) takes over as the predominant form of hemoglobin in normal children.

Certain genetic abnormalities can cause the switch to adult hemoglobin synthesis to fail, resulting in a condition known as hereditary persistence of fetal hemoglobin (HPFH).

Structure and Genetics

Most types of normal hemoglobin, including hemoglobin A, hemoglobin A2, as well as hemoglobin F, are tetramers composed of four protein subunits and four heme prosthetic groups. Whereas adult hemoglobin is composed of two á (alpha) and two â (beta) subunits, fetal hemoglobin is composed of two á subunits and two ã (gamma) subunits, and is commonly denoted as $á_2ã_2$. Because of its presence in fetal hemoglobin, the ã subunit is commonly called the "fetal" hemoglobin subunit.

In humans, the gamma subunit is encoded on chromosome 11, as is the beta subunit. There are two similar copies of the gamma subunit gene: ãG which has a glycine at position 136, and ãA which has an alanine. The gene that codes for the alpha subunit is located on chromosome 16 and is also present in duplicate.

Clinical Significance

Treatment of Sickle-Cell Disease

When fetal hemoglobin production is switched off after birth, normal children begin producing adult hemoglobin (HbA). Children with sickle-cell disease instead begin producing a defective form of hemoglobin called hemoglobin S which aggregates together and forms filaments that cause red blood cells to change their shape from round to sickle-shaped. These defective red blood cells have a greater tendency to stack on top of one another and block blood vessels. These invariably lead to so-called painful vaso-occlusive episodes, which are a hallmark of the disease.

If fetal hemoglobin remains the predominant form of hemoglobin after birth, the number of painful episodes decreases in patients with sickle-cell disease. Hydroxyurea promotes the production of fetal hemoglobin and can thus be used to treat sickle-cell disease. The fetal hemoglobin's reduction in the severity of the disease comes from its ability to inhibit the formation of hemoglobin aggregates within red blood cells which also containing hemoglobin S. Combination therapy with hydroxyurea and recombinant erythropoietin—rather than treatment with hydroxyurea alone—has been shown to further elevate hemoglobin F levels and to promote the development of HbF-containing F-cells.

HEMOGLOBIN PROTEIN SUBUNITS (GENES)

HEMOGLOBIN, ALPHA 1

Hemoglobin, alpha 1, also known as HBA1, is a hemoglobin protein that in humans is encoded by the HBA1 gene. The human alpha globin gene cluster located on chromosome 16 spans about 30 kb and includes seven loci: 5'- zeta - pseudozeta - mu - pseudoalpha-1 - alpha-2 - alpha-1 - theta - 3'. The alpha-2 (HBA2) and alpha-1 (HBA1; this gene) coding sequences are identical. These genes differ slightly over the 5' untranslated regions and the introns, but they differ significantly over the 3' untranslated regions.

- *Protein:* Two alpha chains plus two beta chains constitute HbA, which in normal adult life comprises about 97% of the total hemoglobin; alpha chains combine with delta chains to constitute HbA-2, which with HbF (fetal hemoglobin) makes up the remaining 3% of adult hemoglobin.
- *Clinical significance:* Alpha thalassemias result from deletions of each of the alpha genes as well as deletions of both HBA2 and HBA1; some nondeletion alpha thalassemias have also been reported.
- *Interactions:* Hemoglobin, alpha 1 has been shown to interact with HBB.

HBB

Beta globin (also referred to as HBB, â-globin, haemoglobin beta, hemoglobin beta, or preferably haemoglobin subunit beta) is a globin protein, which along with alpha globin (HBA), makes up the most common form of haemoglobin in adult humans, the HbA. It is 146 amino acids long and has a molecular weight of 15,867 Da. Normal adult human HbA is a heterotetramer consisting of two alpha chains and two beta chains. HBB is encoded by the HBB gene on human chromosome 11. Mutations in the gene produce several variants of the proteins which are implicated with genetic disorders such as sickle-cell disease and beta thalassemia, as well as beneficial traits such as genetic resistance to malaria.

Gene Locus

HBB protein is produced by the gene HBB which is located in the multigene locus of â-globin locus on chromosome 11, specifically on the short arm position 15.5. Expression of beta globin and the neighbouring globins in the â-globin locus is controlled by single locus control region (LCR), the most important regulatory element in the locus located upstream of the globin genes. The normal allelic variant is 1600 base pairs (bp) long and contains three exons. The order of the genes in the beta-globin cluster is 5' - epsilon – gamma-G – gamma-A – delta – beta - 3'.

Interactions

HBB interacts with Hemoglobin, alpha 1 (HBA1) to form haemoglobin A, the major haemoglobin in adult humans. The interaction is two-fold. First, one HBB and one HBA1 combine to form a single polypetide chain (dimer). Secondly, two dimers combine to form a larger polypeptide (tetramer), and this becomes the functional haemolglobin.

Associated Genetic Disorders

Beta Thalassemia

Total or partial absence of HBB causes a genetic disease called beta thalassemia. Total loss called, thalassemia major or beta-0-thalassemia, is due to mutation in both alleles, and this results in failure to form beta chain of haemoglobin. It prevents oxygen supply in the tissues. It is highly lethal. Symptoms, such as severe anaemia and heart attack, appear within two years after birth. They can be treated only by life-long blood transfusion and bone marrow transplantation. Reduced HBB function called thalassemia minor or beta+ thalassemia is due to mutation in one of the alleles. It is less severe but patients are prone to other diseases such as asthma and liver problems.

Haemoglobin C

Sickle cell disease is closely related to another mutant haemoglobin called haemoglobin C (HbC), because they can be inherited together. HbC mutation is at the same position in HBB, but glutamic acid is replaced by lysine (â6Glu'!Lys). The mutation is particularly prevalent in West African populations. HbC provides near full protection against Plasmodium falciparum in homozygous

(CC) individuals and intermediate protection in heterozygous (AC) individuals. This indicates that HbC has stronger influence than HbS, and is predicted to replace HbS in malaria-endemic regions.

Haemoglobin E

Another point mutation in HBB, in which glutamic acid is replaced with lysine at position 26, leads to the formation of haemoglobin E (HbE). HbE has a very unstable á- and â-globin association. Even though the unstable protein itself has mild effect, inherited with HbS and thalassemia traits, it turns into a life-threatening form of â-thalassemia. The mutation is of relatively recent origin suggesting that it resulted from selective pressure against severe falciparum malaria, as heterozygous allele prevents the development of malaria.

Sickle Cell Disease

More than a thousand naturally occurring HBB variants have been discovered. The most common is HbS, which causes sickle cell disease. HbS is produced by a point mutation in HBB in which the codon GAG is replaced by GTG. This results in the replacement of hydrophilic amino acid glutamic acid with the hydrophobic amino acid valine at the sixth position (β6Glu→ Val). This substitution creates a hydrophobic spot on the outside of the protein that sticks to the hydrophobic region of an adjacent haemoglobin molecule's beta chain. This further causes clumping of HbS molecules into rigid fibers, causing "sickling" of the entire red blood cells in the homozygous (HbS/HbS) condition. Homozygous allele has become one of the deadliest genetic factors. Whereas, people heterozygous for the mutant allele (HbS/HbA) are resistant to malaria and develop minimal effects of the anaemia.

Human Evolution

Malaria due to Plasmodium falciparum is a major selective factor in human evolution. It has influenced mutations in HBB in various degrees resulting in the existence of numerous HBB variants. Some of these mutations are not directly lethal and instead confer resistance to malaria, particularly in Africa where malaria is epidemic. People of African descent have evolved to have higher rates of the mutant HBB because the heterozygous individuals have a misshaped red blood cell that prevent attacks from malarial parasites. Thus, HBB mutants are the sources of positive selection in these regions and are important for their long-term survival. Such selection markers are important for tracing human ancestry and diversification from Africa.

HBD

Hemoglobin subunit delta is a protein that in humans is encoded by the HBD gene.

Function

The delta (HBD) and beta (HBB) genes are normally expressed in the adult: two alpha chains plus two beta chains constitute HbA, which in normal adult life comprises about 97% of the total hemoglobin. Two alpha chains plus two delta chains constitute HbA-2, which with HbF comprises the remaining 3% of adult hemoglobin. Five beta-like globin genes are found within a 45 kb cluster on chromosome 11 in the following order: 5' - epsilon – gamma-G – gamma-A – delta – beta - 3'.

Clinical Significance

Mutations in the delta-globin gene are associated with Delta-thalassemia.

HEMOGLOBIN COMPOUNDS

CARBAMINOHEMOGLOBIN

Carbaminohemoglobin (or Carbaminohaemoglobin, also known as carbhaemoglobin and carbohaemoglobin) is a compound of hemoglobin and carbon dioxide, and is one of the forms in which carbon dioxide exists in the blood. 10% of carbon dioxide is carried in blood this way (85% carried in blood as bicarbonate [hydrogen carbonate], 5% carried as free CO2, in solution).

MECHANISM

When carbon dioxide binds to hemoglobin, carbaminohemoglobin is formed, lowering haemoglobin's affinity for oxygen via the Bohr effect. In the absence of oxygen, unbound hemoglobin molecules have a greater chance of becoming carbaminohaemoglobin. (The Haldane Effect relates to the increased affinity of deoxygenated hemoglobin for H+: offloading of oxygen to the tissues thus results in increased affinity of the hemoglobin for carbon dioxide, and H+ - which the body wants to be rid of- which can then be transported to the lung for removal). Carbaminohemoglobin has a distinctive blue color that may contribute to the dark red color of deoxygenated venous blood (compared to bright, saturated red of oxygenated arterial blood), blue color of veins and the purplish or bluish color of tissues in hypoxia. Hemoglobin can bind to four molecules of carbon dioxide. The carbon dioxide molecules form an amide linkage to the four terminal-amine groups of the four protein chains in the deoxy form of the molecule. Thus, one hemoglobin molecule can transport four carbon dioxide molecules back to the lungs, where they are released when the molecule changes back to the oxyhemoglobin form.

CARBOXYHEMOGLOBIN

Carboxyhemoglobin (British English: Carboxyhaemoglobin) (COHb) is a stable complex of carbon monoxide and hemoglobin that forms in red blood cells upon contact with carbon monoxide (CO).

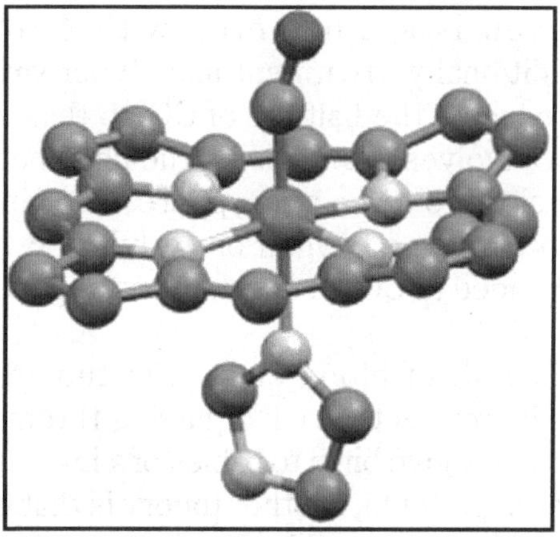

Fig. A Heme Unit of Human Carboxyhaemoglobin, Showing the Carbonyl Ligand at the Apical Position, Trans to the Histidine Residue.

Large quantities of CO hinders the ability of Hb to deliver oxygen to the body. CO is produced in normal metabolism and is also a common chemical. Tobacco smoking (through carbon monoxide inhalation) raises the blood levels of COHb by a factor of several times from its normal concentrations.

Affinity of Hb for CO

Hemoglobin contains four heme groups each capable of reversibly binding to one oxygen molecule. Oxygen binding to any of these sites causes a conformational change in the protein, facilitating binding to each of the other sites. Carbon monoxide binds to hemoglobin at the same sites as oxygen, but approximately 200 times more tightly. Normally, oxygen would bind to hemoglobin in the lungs and be released in areas with low oxygen partial pressure (e.g. active muscles). When carbon monoxide binds to hemoglobin, it cannot be released as easily as oxygen. The slow release rate of carbon monoxide causes an accumulation of CO-bound hemoglobin molecules as exposure to carbon monoxide continues. Because of this, fewer hemoglobin particles are available to bind and deliver oxygen, thus causing the gradual suffocation associated with carbon monoxide poisoning.

Carbon Monoxide as a Poison

Since COHb releases carbon monoxide slowly, less haemoglobin will be available to transport oxygen from the lungs to the rest of the body. Conversion of most Hb to COHb results in death - known medically as carboxyhemoglobinemia or carbon monoxide poisoning. Smaller amounts COHb lead to oxygen deprivation of the body causing tiredness, dizziness, and unconsciousness.

COHb has a half-life in the blood of 4 to 6 hours. This time can be reduced to 70 to 35 minutes with administration of pure oxygen (the lower number applying when oxygen is administered with 4 to 5% CO2 to cause hyperventilation). Additionally, treatment in a Hyperbaric Chamber is a more effective manner of reducing the half-life of COHb than administering oxygen alone. This treatment involves pressurizing the chamber with pure oxygen at an absolute pressure close to three atmospheres allowing the body's fluids to absorb oxygen and to pass free oxygen on to hypoxic tissues instead of the crippled hemoglobin bonded to CO. In effect, bypassing the need for hemoglobin in the blood.

COHb increases risk of blood clot. It is thought that through this mechanism, smoking increases the risk of having thromboembolic disease.

Pregnant smokers may give birth to babies of a lower birth mass. In addition to vasoconstriction of the placenta, another theory is that since fetal hemoglobin takes up carbon monoxide more readily than in an adult, therefore the fetus of a smoker will suffer from mild hypoxia, potentially retarding its development.

Carbon Monoxide as Normal Biological Metabolite and Potential Pharmaceutical

In biology, carbon monoxide is naturally produced by the action of heme oxygenase 1 and 2 on the heme from hemoglobin breakdown. This process produces a certain amount of carboxyhemoglobin in normal persons, even if they do not breathe any carbon monoxide.

Following the first report that carbon monoxide is a normal neurotransmitter in 1993, as well as one of three gasses which naturally modulate inflammatory responses in the body (the other two being nitric oxide and hydrogen sulfide), carbon monoxide has received a great deal of clinical attention as a biological regulator. In many tissues, all three gases are known to act as anti-inflammatories, vasodilators and encouragers of neovascular growth. However, the issues are complex, as neovascular growth is not always beneficial, since it plays a role in tumor growth, and also the damage from wet macular degeneration, a disease for which smoking (a major source of carbon monoxide in the blood, several times more than natural production) increases the risk from 4 to 6 times. Studies involving carbon monoxide have been conducted in many laboratories throughout the world for its anti-inflammatory and cytoprotective properties. These properties could potentially be used to prevent the development of a series of pathological conditions including ischemia reperfusion injury, transplant rejection, atherosclerosis, severe sepsis, severe malaria or autoimmunity. Clinical tests involving humans have been performed. As of 2009, however, the results had not yet been released.

VASKA'S COMPLEX

Vaska's complex is the trivial name for the chemical compound trans-chlorocarbonylbis(triphenylphosphine)iridium(I), which has the formula

$IrCl(CO)[P(C_6H_5)_3]_2$. This square planar diamagnetic organometallic complex consists of a central iridium atom bound to two mutually trans triphenylphosphine ligands, carbon monoxide, and a chloride ion. The complex was first reported by J. W. DiLuzio and Lauri Vaska in 1961. Vaska's complex can undergo oxidative addition and is notable for its ability to bind to O_2 reversibly. It is a bright yellow crystalline solid.

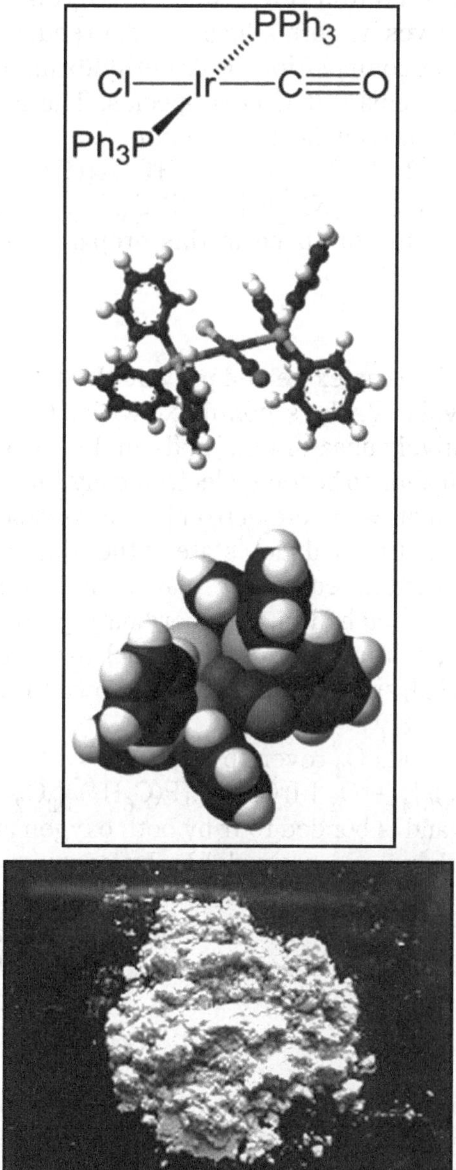

Fig. Vaska's Complex.

Preparation

The synthesis involves heating virtually any iridium chloride salt with triphenylphosphine and a carbon monoxide source. The most popular method uses dimethylformamide (DMF) as a solvent, and sometimes aniline is added to accelerate the reaction. Another popular solvent is 2-methoxyethanol. The reaction is typically conducted under nitrogen. In the synthesis, triphenylphosphine serves as both a ligand and a reductant, and the carbonyl ligand is derived by decomposition of dimethylformamide, probably via a deinsertion of an intermediate Ir-C(O)H species. The following is a possible balanced equation for this complicated reaction.

$IrCl_3(H_2O)_3$ + 3 $P(C_6H_5)_3$ + $HCON(CH_3)_2$ + $C_6H_5NH_2$! $IrCl(CO)[P(C_6H_5)_3]_2$ + $[(CH_3)_2NH_2]Cl$ + $OP(C_6H_5)_3$ + $[C_6H_5NH_3]Cl$ + 2 H_2O

Typical sources of iridium used in this preparation are $IrCl_3.xH_2O$ and H_2IrCl_6.

Reactions

Studies on Vaska's complex helped provide the conceptual framework for homogeneous catalysis. Vaska's complex, with 16 valence electrons, is considered "coordinatively unsaturated" and can thus bind to one two-electron or two one-electron ligands to become electronically saturated with 18 valence electrons. The addition of two one-electron ligands is called oxidative addition. Upon oxidative addition, the oxidation state of the iridium increases from Ir(I) to Ir(III). The four-coordinated square planar arrangement in the starting complex converts to an octahedral, six-coordinate product. Vaska's complex undergoes oxidative addition with conventional oxidants such as halogens, strong acids such as HCl, and other molecules known to react as electrophiles, such as iodomethane (CH_3I).

Vaska's complex binds O_2 reversibly:

$IrCl(CO)[P(C_6H_5)_3]_2$ + O_2 ! $IrCl(CO)[P(C_6H_5)_3]_2O_2$

The dioxygen ligand is bonded to Ir by both oxygen atoms, so-called side-on bonding. In myoglobin and hemoglobin, by contrast, O_2 binds "end-on," attaching to the metal via only one of the two oxygen atoms. The resulting dioxygen adduct reverts to the parent complex upon heating or purging the solution with an inert gas, signaled by a colour change from orange back to yellow.

Spectroscopy

Infrared spectroscopy can be used to analyse the products of oxidative addition to Vaska's complex because the reactions induce characteristic shifts of the stretching frequency of the coordinated carbon monoxide. These shifts are dependent on the amount of π-back bonding allowed by the newly associated ligands.

The CO stretching frequencies for Vaska's complex and oxidatively added ligands have been documented in the literature.

- Vaska's Complex: 1967 cm^{-1}
- Vaska's + O_2: 2015 cm^{-1}
- Vaska's + MeI: 2047 cm^{-1}
- Vaska's + I_2: 2067 cm^{-1}

Oxidative addition to give Ir(III) products reduces the ð-bonding from Ir to C, which causes the increase in the frequency of the carbonyl stretching band. The stretching frequency change depends upon the ligands that have been added, but the frequency is always greater than 2000 cm"1 for an Ir(III) complex.

CHLOROPHYLL

Chlorophyll (also chlorophyl) is a term used for several closely related green pigments found in cyanobacteria and the chloroplasts of algae and plants. Its name is derived from the Greek words ÷ëùñüò, chloros ("green") and öýëëïí, phyllon ("leaf"). Chlorophyll is an extremely important biomolecule, critical in photosynthesis, which allows plants to absorb energy from light. Chlorophyll absorbs light most strongly in the blue portion of the electromagnetic spectrum, followed by the red portion. Conversely, it is a poor absorber of green and near-green portions of the spectrum, hence the green color of chlorophyll-containing tissues. Chlorophyll was first isolated and named by Joseph Bienaimé Caventou and Pierre Joseph Pelletier in 1817.

Chlorophyll and Photosynthesis

Chlorophyll is vital for photosynthesis, which allows plants to absorb energy from light. Chlorophyll molecules are specifically arranged in and around photosystems that are embedded in the thylakoid membranes of chloroplasts. In these complexes, chlorophyll serves two primary functions. The function of the vast majority of chlorophyll (up to several hundred molecules per photosystem) is to absorb light and transfer that light energy by resonance energy transfer to a specific chlorophyll pair in the reaction center of the photosystems.

The two currently accepted photosystem units are Photosystem II and Photosystem I, which have their own distinct reaction centres, named P680 and P700, respectively. These centres are named after the wavelength (in nanometers) of their red-peak absorption maximum. The identity, function and spectral properties of the types of chlorophyll in each photosystem are distinct and determined by each other and the protein structure surrounding them. Once extracted from the protein into a solvent (such as acetone or methanol), these chlorophyll pigments can be separated in a simple paper chromatography experiment and, based on the number of polar groups between chlorophyll a and chlorophyll b, will chemically separate out on the paper. The function of the reaction center chlorophyll is to use the energy absorbed by and transferred to it from the other chlorophyll pigments in the photosystems to undergo a charge separation, a specific redox reaction in which the chlorophyll donates an electron into a series of molecular intermediates called an electron transport chain. The charged reaction center chlorophyll ($P680^+$) is then reduced back to its ground state by accepting an electron.

In Photosystem II, the electron that reduces $P680^+$ ultimately comes from the oxidation of water into O_2 and H^+ through several intermediates. This reaction is how photosynthetic organisms such as plants produce O_2 gas, and is the source for practically all the O_2 in Earth's atmosphere. Photosystem I typically works in series with Photosystem II; thus the $P700^+$ of Photosystem I is usually reduced, via many intermediates in the thylakoid membrane, by electrons ultimately from Photosystem II. Electron transfer reactions in the thylakoid membranes are complex, however, and the source of electrons used to reduce $P700^+$ can vary.

The electron flow produced by the reaction center chlorophyll pigments is used to shuttle H^+ ions across the thylakoid membrane, setting up a chemiosmotic potential used mainly to produce ATP chemical energy; and those electrons ultimately reduce $NADP^+$ to NADPH, a universal reductant used to reduce CO_2 into sugars as well as for other biosynthetic reductions.

Reaction center chlorophyll–protein complexes are capable of directly absorbing light and performing charge separation events without other chlorophyll pigments, but the absorption cross section (the likelihood of absorbing a photon under a given light intensity) is small. Thus, the remaining chlorophylls in the photosystem and antenna pigment protein complexes associated with the photosystems all cooperatively absorb and funnel light energy to the reaction center. Besides chlorophyll a, there are other pigments, called accessory pigments, which occur in these pigment–protein antenna complexes.

Chemical Structure

Chlorophyll is a chlorin pigment, which is structurally similar to and produced through the same metabolic pathway as other porphyrin pigments such as heme. At the center of the chlorin ring is a magnesium ion. This was discovered in 1906, and was the first time that magnesium had been detected in living tissue. For the structures depicted in this article, some of the ligands attached to the Mg^{2+} center are omitted for clarity. The chlorin ring can have several different side chains, usually including a long phytol chain. There are a few different forms that occur naturally, but the most widely distributed form in terrestrial plants is chlorophyll a. After initial work done by German chemist Richard Willstätter spanning from 1905 to 1915, the general structure of chlorophyll a was elucidated by Hans Fischer in 1940. By 1960, when most of the stereochemistry of chlorophyll a was known, Robert Burns Woodward published a total synthesis of the molecule. In 1967, the last remaining stereochemical elucidation was completed by Ian Fleming, and in 1990 Woodward and co-authors published an updated synthesis. Chlorophyll f was announced to be present in cyanobacteria and other oxygenic microorganisms that form stromatolites in 2010; a molecular formula of $C_{55}H_{70}O_6N_4Mg$ and a structure of (2-formyl)-chlorophyll a were deduced based on NMR, optical and mass spectra. The different structures of chlorophyll are summarized below:

	Chlorophyll a	**Chlorophyll b**	**Chlorophyll c1**	**Chlorophyll c2**	**Chlorophyll d**	**Chlorophyll f**
Molecular formula	$C_{55}H_{72}O_5N_4Mg$	$C_{55}H_{70}O_6N_4Mg$	$C_{35}H_{30}O_5N_4Mg$	$C_{35}H_{28}O_5N_4Mg$	$C_{54}H_{70}O_6N_4Mg$	$C_{55}H_{70}O_6N_4Mg$
C2 group	$-CH_3$	$-CH_3$	$-CH_3$	$-CH_3$	$-CH_3$	-CHO
C3 group	$-CH{=}CH_2$	$-CH{=}CH_2$	$-CH{=}CH_2$	$-CH{=}CH_2$	-CHO	$-CH{=}CH_2$
C7 group	$-CH_3$	-CHO	$-CH_3$	$-CH_3$	$-CH_3$	$-CH_3$
C8 group	$-CH_2CH_3$	$-CH_2CH_3$	$-CH_2CH_3$	$-CH{=}CH_2$	$-CH_2CH_3$	$-CH_2CH_3$
C17 group	$-CH_2CH_2COO$-Phytyl	$-CH_2CH_2COO$-Phytyl	$-CH{=}CHCOOH$	$-CH{=}CHCOOH$	$-CH_2CH_2COO$-Phytyl	$-CH_2CH_2COO$-Phytyl
C17-C18 bond	Single (chlorin)	Single (chlorin)	Double (porphyrin)	Double (porphyrin)	Single (chlorin)	Single (chlorin)
Occurrence	Universal	Mostly plants	Various algae	Various algae	Cyanobacteria	Cyanobacteria

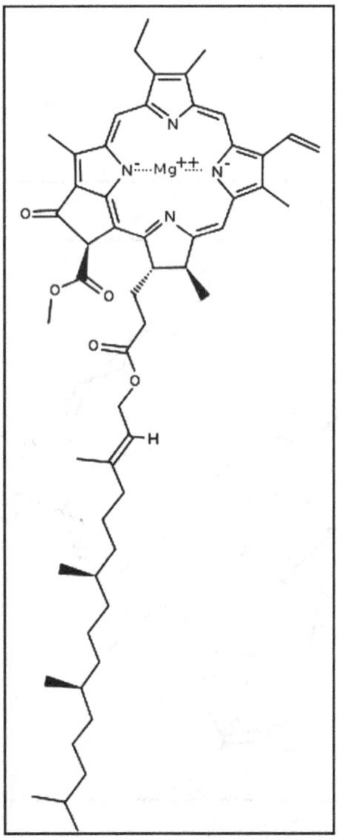

Structure of chlorophyll a

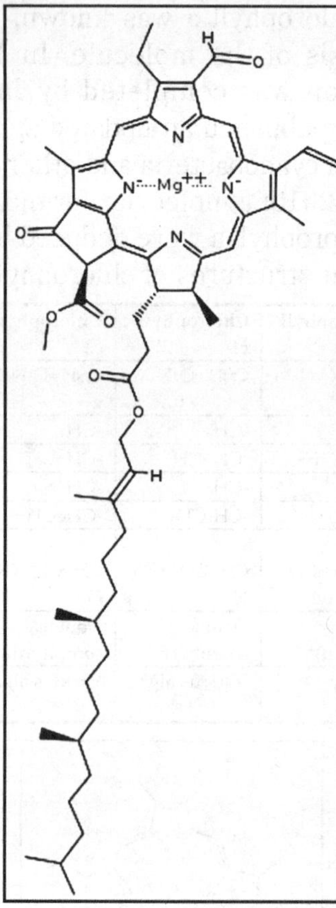

Structure of chlorophyll b

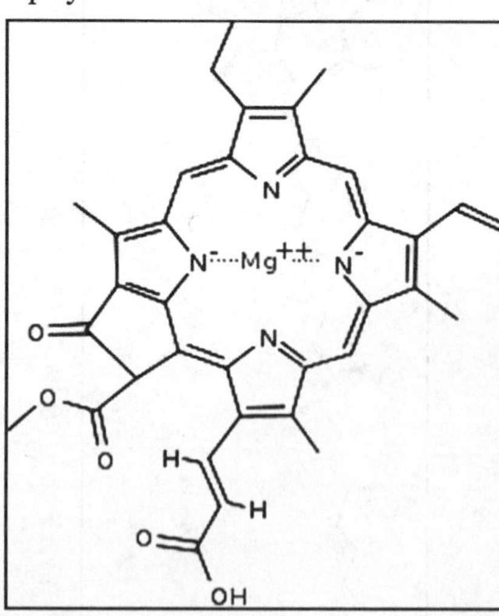

Structure of chlorophyll c1

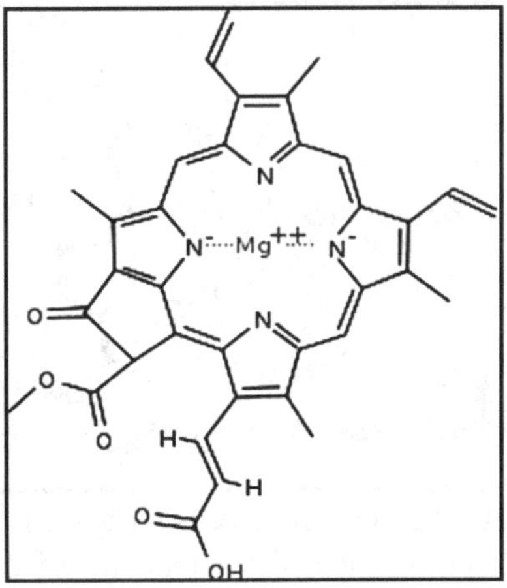

Structure of chlorophyll c2

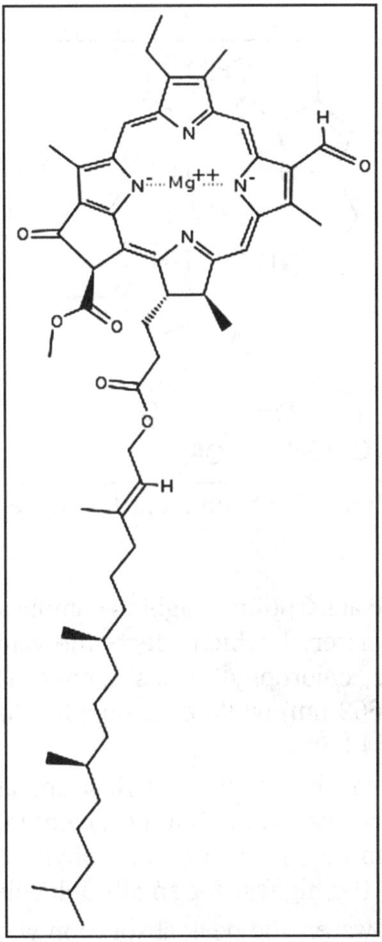

Structure of chlorophyll d

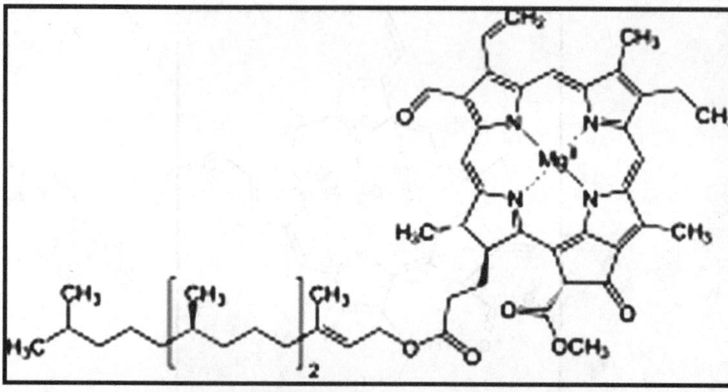

Structure of chlorophyll fWhen leaves degreen in the process of plant senescence, chlorophyll is converted to a group of colourless tetrapyrroles known as nonfluorescent chlorophyll catabolites (NCC's) with the general structure:

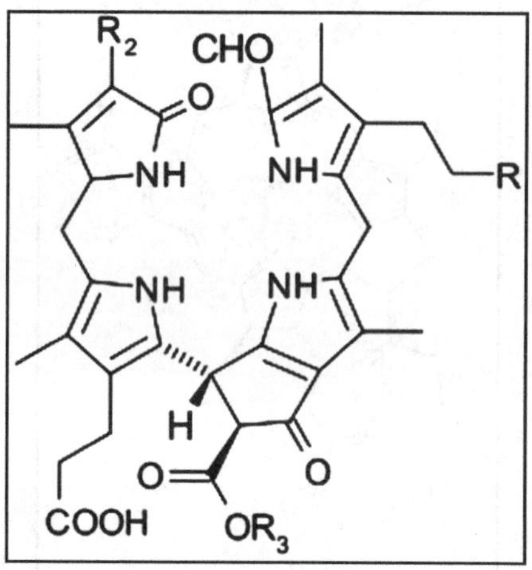

These compounds have also been identified in several ripening fruits.

Spectrophotometry

Measurement of the absorption of light is complicated by the solvent used to extract it from plant material, which affects the values obtained,

- In diethyl ether, chlorophyll a has approximate absorbance maxima of 430 nm and 662 nm, while chlorophyll b has approximate maxima of 453 nm and 642 nm.
- The absorption peaks of chlorophyll a are at 665 nm and 465 nm. Chlorophyll a fluoresces at 673 nm (maximum) and 726 nm. The peak molar absorption coefficient of chlorophyll a exceeds 10^5 M"1 cm"1, which is among the highest for small-molecule organic compounds.
- In 90% acetone-water, the peak absorption wavelengths of chlorophyll a are 430 nm and 664 nm; peaks for chlorophyll b are 460 nm and

647 nm; peaks for chlorophyll c1 are 442 nm and 630 nm; peaks for chlorophyll c2 are 444 nm and 630 nm; peaks for chlorophyll d are 401 nm, 455 nm and 696 nm.

By measuring the absorption of light in the red and far red regions it is possible to estimate the concentration of chlorophyll within a leaf.In his scientific paper Gitelson (1999) states, "The ratio between chlorophyll fluorescence, at 735 nm and the wavelength range 700nm to 710 nm, F735/F700 was found to be linearly proportional to the chlorophyll content (with determination coefficient, r2, more than 0.95) and thus this ratio can be used as a precise indicator of chlorophyll content in plant leaves." The fluorescent ratio chlorophyll content meters use this technique.

Biosynthesis

In plants, chlorophyll may be synthesized from succinyl-CoA and glycine, although the immediate precursor to chlorophyll a and b is protochlorophyllide. In Angiosperm plants, the last step, conversion of protochlorophyllide to chlorophyll, is light-dependent and such plants are pale (etiolated) if grown in the darkness. Non-vascular plants and green algae have an additional light-independent enzyme and grow green in the darkness instead.

Chlorophyll itself is bound to proteins and can transfer the absorbed energy in the required direction. Protochlorophyllide occurs mostly in the free form and, under light conditions, acts as a photosensitizer, forming highly toxic free radicals. Hence, plants need an efficient mechanism of regulating the amount of chlorophyll precursor. In angiosperms, this is done at the step of aminolevulinic acid (ALA), one of the intermediate compounds in the biosynthesis pathway. Plants that are fed by ALA accumulate high and toxic levels of protochlorophyllide; so do the mutants with the damaged regulatory system. Chlorosis is a condition in which leaves produce insufficient chlorophyll, turning them yellow. Chlorosis can be caused by a nutrient deficiency of iron—called iron chlorosis—or by a shortage of magnesium or nitrogen. Soil pH sometimes plays a role in nutrient-caused chlorosis; many plants are adapted to grow in soils with specific pH levels and their ability to absorb nutrients from the soil can be dependent on this. Chlorosis can also be caused by pathogens including viruses, bacteria and fungal infections, or sap-sucking insects.

Complementary Light Absorbance of Anthocyanins with Chlorophylls

Anthocyanins are other plant pigments. The absorbance pattern responsible for the red color of anthocyanins may be complementary to that of green chlorophyll in photosynthetically active tissues such as young Quercus coccifera leaves. It may protect the leaves from attacks by plant eaters that may be attracted by green color.

Culinary use

Chlorophyll is registered as a food additive (colorant), and its E number is E140. Chefs use chlorophyll to color a variety of foods and beverages green, such as pasta and absinthe. Chlorophyll is not soluble in water, and it is first mixed with a small quantity of vegetable oil to obtain the desired solution. Extracted liquid chlorophyll was considered to be unstable and always denatured until 1997, when Frank S. & Lisa Sagliano used freeze-drying of liquid chlorophyll at the University of Florida and stabilized it as a powder, preserving it for future use.

Alternative Medicine

Many claims are made about the healing properties of chlorophyll, but most have been disproved or are exaggerated by the companies that are marketing them. Quackwatch, a website dedicated to debunking false medical claims, has a quote from Toledo Blade (1952) which claims "Chlorophyll Held Useless As Body Deodorant", but later has John C. Kephart pointing out "No deodorant effect can possibly occur from the quantities of chlorophyll put in products such as gum, foot powder, cough drops, etc. To be effective, large doses must be given internally".

GLOBIN FOLD

The globin fold is a common three-dimensional fold in proteins and defines the globin-like protein superfamily. This fold typically consists of eight alpha helices, although some proteins have additional helix extensions at their termini. The globin fold is found in its namesake globin protein families: hemoglobins and myoglobins, as well as in phycocyanins. Because myoglobin was the first protein whose structure was solved, the globin fold was thus the first protein fold discovered. Since the globin fold contains only helices, it is classified as an all-alpha protein fold.

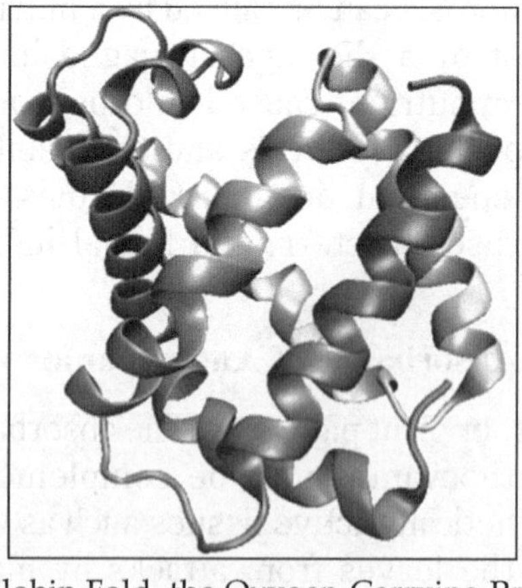

Fig. An Example of the Globin Fold, the Oxygen-Carrying Protein Myoglobin (PDB ID 1MBA) from the Mollusc Aplysia Limacina.

Helix Packing

The eight helices of the globin fold core share significant nonlocal structure, unlike other structural motifs in which amino acids close to each other in primary sequence are also close in space. The helices pack together at an average angle of about 50 degrees, significantly steeper than other helical packings such as the helix bundle. The exact angle of helix packing depends on the sequence of the protein, because packing is mediated by the sterics and hydrophobic interactions of the amino acid side chains near the helix interfaces.

Sequence Conservation

Although the globin fold is highly evolutionarily conserved, the sequences that form the fold can have as low as 16% sequence identity. While the sequence specificity of the fold is not stringent, the hydrophobic core of the protein must be maintained and hydrophobic patches on the generally hydrophilic solvent-exposed surface must be avoided in order for the structure to remain stable and soluble. The most famous mutation in the globin fold is a change from glutamate to valine in one chain of the hemoglobin molecule. This mutation creates a "hydrophobic patch" on the protein surface that promotes intermolecular aggregation, the molecular event that gives rise to sickle-cell anemia.

HEMEPROTEIN

A hemeprotein (or haemprotein; also hemoprotein or haemoprotein), or heme protein, is a metalloprotein containing a heme prosthetic group- an organic compound that allows a protein to carry out several functions that it cannot do alone. Heme remains bound to the protein permanently, either covalently or noncovalently bound or both.

The heme contains a reduced iron atom, Fe2+ in the center of a highly hydrophobic, planar, porphyrin ring. The iron has six possible coordination bonds. The porphyrin ring has 4 nitrogen atoms that bind to the iron, leaving two other coordination positions of the iron available for bonding to the histidine of the protein and a divalent atom.

Hemeproteins probably evolved to incorporate the iron (Fe) atom contained within the protoporphyrin IX ring of heme into proteins. As it makes hemeproteins responsive to molecules that can bind divalent iron (Fe), this strategy has been maintained throughout evolution as it plays crucial physiological functions. Oxygen (O_2) nitric oxide (NO), carbon monoxide (CO) and hydrogen sulfide (H2S) bind to the iron atom in heme proteins. Once bound to the prosthetic heme groups, these molecules can modulate the activity/function of those hemeproteins, affording signal transduction. Therefore, when produced in biologic systems (cells), these gaseous molecules are referred to as gasotransmitters.

Because of their diverse range of biological functions, the structural and functional characterizations of hemeproteins are the most studied classes of biomolecules. Data on heme protein structure and function has been aggregated into The Heme Protein Database (HPD), a secondary database to the Protein Data Bank.

Roles

Hemeproteins have diverse biological functions including oxygen transport, which is completed via hemeproteins including hemoglobin, myoglobin, neuroglobin, cytoglobin and leghemoglobin. Catalysis occurs with hemeproteins cytochrome P450s, cytochrome c oxidase, ligninases and peroxidases. Hemeproteins also enable electron transfer as they form part of the electron transport chain. Cytochrome a, cytochrome b and cytochrome c are responsible for electron transfer functions. The sensory system also relies on some hemeproteins including FixL, an Oxygen sensor, CooA, a carbon monoxide sensor, and soluble guanylyl cyclase. Finally, the hemeprotein catalase aids in biological defense.

Hemoglobin and Myoglobin

Hemoglobin and Myoglobin are examples of hemeprotein that are essential in the storing and transports of oxygen in mammals. Hemoglobin is a quaternary protein that occurs in the red blood cell, whereas, myoglobin is a tertiary protein found the muscle cells of mammals. Although they might differ in location and size, their function are similar. Being hemeproteins, they both contain a prosthetic group.

His-F8 of the myoglobin, also known as the proximal histidine is covalently bonded to the 5th coordination position of the iron. Oxygen interacts with the distal His by ways of a hydrogen bond,not a covalent one. It binds to the 6th coordination position of the iron, His-E7 of the myoglobin binds to the Oxygen that is now covalently bonded to the iron. The same occurs for the hemoglobin, however, being a protein with 4 subunits, the hemoglobin contains 4 heme, therefore, allowing 4 Oxygen atoms to bind to the heme. Myoglobin and hemoglobin are globular proteins that serve to bind and deliver oxygen. These globins dramatically improve the concentration of molecular oxygen that can be dissolved in the biological fluids of vertebrates and some invertebrates.

Both of these globins contain an organic prosthetic group for binding oxygen and they contain many similar structural characteristics. Yet the differences in function between these two proteins can be explored from a structural perspective, leading to a discourse on ligand binding and allosteric regulation.

Myoglobin and hemoglobin are of obvious critical importance to vertebrate life, yet they are also studied because they serve as excellent, well-characterized models for illustrating the principles of protein structure, dynamics, and function.

Myoglobin

Myoglobin is found in vertebrate muscle cells. Muscle cells, when put into action, can quickly require a large amount of oxygen for respiration because of their incredible demand for energy. Therefore, muscle cells use myoglobin to accelerate oxygen diffusion and act as localized oxygen reserves for times of intense respiration.

Hemoglobin

In vertebrates, hemoglobin is found in the cytosol of red blood cells in the bloodstream. Hemoglobin is sometimes referred to as the oxygen transport protein, in order to contrast it with its stationary cousin myoglobin, although its function and mechanism are more complex than this name would suggest.

In vertebrates, oxygen is taken into the body by the tissues of the lungs, and passed to the red blood cells in the bloodstream. Oxygen is then distributed to all of the tissues in the body and offloaded from the red blood cells to respiring cells. Hemoglobin then picks up carbon dioxide to be returned to the lungs. Thus, hemoglobin binds and offloads both oxygen and carbon dioxide at the appropriate tissues, serving to deliver the oxygen needed for cellular metabolism and removing the resulting waste product, CO_2.

Cytochrome c Oxidase

Cytochrome c Oxidase is an enzyme embedded in the inner membrane of mitochondria. Its main function is to oxidise the Cytochrome c protein. Cytochrome c oxidase contains several metal active sites.

4

The Anemia

Anemia or anaemia is usually defined as a decrease in the amount of red blood cells (RBCs) or hemoglobin in the blood. It can also be defined as a lowered ability of the blood to carry oxygen. When anemia comes on slowly the symptoms are often vague and may include: feeling tired, weakness, shortness of breath or a poor ability to exercise. Anemia that comes on quickly often has greater symptoms which may include: confusion, feeling like one is going to pass out, and increased thirst. Anemia must be significant before a person becomes noticeably pale. Additional symptoms may occur depending on the underlying cause.

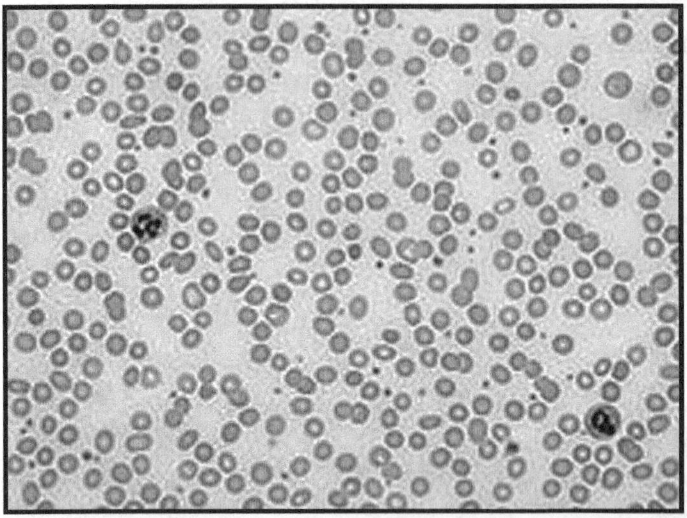

Fig. Human Blood from a Case of Iron-Deficiency Anemia.

There are three main types of anemia: that due to blood loss, that due to decreased red blood cell production, and that due to increased red blood cell breakdown. Causes of blood loss include trauma and gastrointestinal bleeding, among others. Causes of decreased production include iron deficiency, a lack of vitamin B12, thalassemia and a number of neoplasms of the bone marrow among others. Causes of increased breakdown include a number of genetic conditions such as sickle cell anemia, infections like malaria and some autoimmune diseases among others. It can also be classified based on the size of red blood cells and amount of hemoglobin in each cell. If the cells are small it is microcytic anemia, if they are large it is macrocytic anemia and if they are normal sized it is normocytic anemia. Diagnosis in men is based on a hemoglobin of less than 130 to 140 g/L (13 to 14 g/dL), while in women it must be less than 120 to 130 g/L (12 to 13 g/dL). Further testing is then required to determine the cause.

Certain groups of individuals, such as pregnant women, benefit from the use of iron pills for prevention. Dietary supplementation, without determining the specific cause, is not recommended. The use of blood transfusions is typically based on a person's signs and symptoms. In those without symptoms they are not recommended unless hemoglobin levels are less than 60 to 80 g/L (6 to 8 g/dL). These recommendations may also apply to some people with acute bleeding. Erythropoiesis-stimulating medications are only recommended in those with severe anemia.

Anemia is the most common disorder of the blood with it affecting about a quarter of people globally. Iron-deficiency anemia affects nearly 1 billion. In 2013 anemia due to iron deficiency resulted in about 183,000 deaths – down from 213,000 deaths in 1990. It is more common in females than males, among children, during pregnancy, and in the elderly. Anemia increases costs of medical care and lowers a person's productivity through a decreased ability to work. The name is derived from Ancient Greek: ναιμια anaimia, meaning "lack of blood", from ν- an-, "not" + αμα haima, "blood".

SIGNS AND SYMPTOMS

Anemia goes undetected in many people, and symptoms can be minor or vague. The signs and symptoms can be related to the underlying cause or the anemia itself. Most commonly, people with anemia report feelings of weakness, or fatigue, general malaise, and sometimes poor concentration. They may also report dyspnea (shortness of breath) on exertion. In very severe anemia, the body may compensate for the lack of oxygen-carrying capability of the blood by increasing cardiac output.

The patient may have symptoms related to this, such as palpitations, angina (if pre-existing heart disease is present), intermittent claudication of the legs, and symptoms of heart failure. On examination, the signs exhibited may include pallor (pale skin, lining mucosa, conjunctiva and nail beds), but this is not a reliable sign. There may be signs of specific causes of anemia, e.g., koilonychia (in iron deficiency), jaundice (when anemia results from abnormal break down of red blood cells — in hemolytic anemia), bone deformities (found in thalassemia major) or leg ulcers. In severe anemia, there may be signs of a hyperdynamic circulation: tachycardia (a fast heart rate), bounding pulse, flow murmurs, and cardiac ventricular hypertrophy (enlargement). There may be signs of heart failure. Pica, the consumption of non-food items such as ice, but also paper, wax, or grass, and even hair or dirt, may be a symptom of iron deficiency, although it occurs often in those who have normal levels of hemoglobin. Chronic anemia may result in behavioral disturbances in children as a direct result of impaired neurological development in infants, and reduced scholastic performance in children of school age. Restless legs syndrome is more common in those with iron-deficiency anemia.

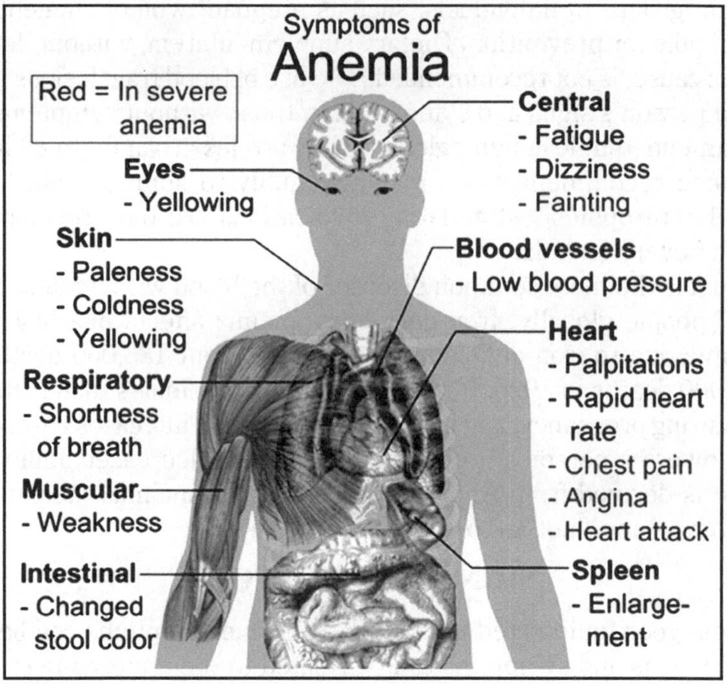

Fig. Main Symptoms that May Appear in Anemia.

CAUSES

Broadly, causes of anemia may be classified as impaired red blood cell (RBC) production, increased RBC destruction (hemolytic anemias), blood loss and fluid overload (hypervolemia). Several of these may interplay to cause anemia eventually. Indeed, the most common cause of anemia is blood loss, but this usually does not cause any lasting symptoms unless a relatively impaired RBC production develops, in turn most commonly by iron deficiency.

IMPAIRED PRODUCTION

- Disturbance of proliferation and differentiation of stem cells
- Pure red cell aplasia

- Aplastic anemia affects all kinds of blood cells. Fanconi anemia is a hereditary disorder or defect featuring aplastic anemia and various other abnormalities.
- Anemia of renal failure by insufficient erythropoietin production
- Anemia of endocrine disorders
- Disturbance of proliferation and maturation of erythroblasts
  - Pernicious anemia is a form of megaloblastic anemia due to vitamin B_{12} deficiency dependent on impaired absorption of vitamin B_{12}. Lack of dietary B_{12} causes non-pernicious megaloblastic anemia
  - Anemia of folic acid deficiency, as with vitamin B_{12}, causes megaloblastic anemia
  - Anemia of prematurity, by diminished erythropoietin response to declining hematocrit levels, combined with blood loss from laboratory testing, generally occurs in premature infants at two to six weeks of age.
  - Iron deficiency anemia, resulting in deficient heme synthesis
  - Thalassemias, causing deficient globin synthesis
  - Congenital dyserythropoietic anemias, causing ineffective erythropoiesis
  - Anemia of renal failure (also causing stem cell dysfunction)
- Other mechanisms of impaired RBC production
  - Myelophthisic anemia or myelophthisis is a severe type of anemia resulting from the replacement of bone marrow by other materials, such as malignant tumors or granulomas.
  - Myelodysplastic syndrome
  - Anemia of chronic inflammation

INCREASED DESTRUCTION

Anemias of increased red blood cell destruction are generally classified as hemolytic anemias. These are generally featuring jaundice and elevated lactate dehydrogenase levels.

- Intrinsic (intracorpuscular) abnormalities cause premature destruction. All of these, except paroxysmal nocturnal hemoglobinuria, are hereditary genetic disorders.
  - Hereditary spherocytosis is a hereditary defect that results in defects in the RBC cell membrane, causing the erythrocytes to be sequestered and destroyed by the spleen.
  - Hereditary elliptocytosis is another defect in membrane skeleton proteins.

- Abetalipoproteinemia, causing defects in membrane lipids
- Enzyme deficiencies
- Pyruvate kinase and hexokinase deficiencies, causing defect glycolysis
- Glucose-6-phosphate dehydrogenase deficiency and glutathione synthetase deficiency, causing increased oxidative stress
- Hemoglobinopathies
- Sickle cell anemia
- Hemoglobinopathies causing unstable hemoglobins
- Paroxysmal nocturnal hemoglobinuria

• Extrinsic (extracorpuscular) abnormalities
  - Antibody-mediated
  - Warm autoimmune hemolytic anemia is caused by autoimmune attack against red blood cells, primarily by IgG. It is the most common of the autoimmune hemolytic diseases. It can be idiopathic, that is, without any known cause, drug-associated or secondary to another disease such as systemic lupus erythematosus, or a malignancy, such as chronic lymphocytic leukemia.
  - Cold agglutinin hemolytic anemia is primarily mediated by IgM. It can be idiopathic or result from an underlying condition.
  - Rh disease, one of the causes of hemolytic disease of the newborn
  - Transfusion reaction to blood transfusions
  - Mechanical trauma to red cells
  - Microangiopathic hemolytic anemias, including thrombotic thrombocytopenic purpura and disseminated intravascular coagulation
  - Infections, including malaria
  - Heart surgery
  - Haemodialysis

BLOOD LOSS

- Anemia of prematurity from frequent blood sampling for laboratory testing, combined with insufficient RBC production
- Trauma or surgery, causing acute blood loss
- Gastrointestinal tract lesions, causing either acute bleeds (e.g. variceal lesions, peptic ulcers or chronic blood loss (e.g. angiodysplasia)
- Gynecologic disturbances, also generally causing chronic blood loss
- From menstruation, mostly among young women or older women who have fibroids

- Infection by intestinal nematodes feeding on blood, such as hookworms and the whipworm Trichuris trichiura.

FLUID OVERLOAD

Fluid overload (hypervolemia) causes decreased hemoglobin concentration and apparent anemia:

- General causes of hypervolemia include excessive sodium or fluid intake, sodium or water retention and fluid shift into the intravascular space.
- Anemia of pregnancy is induced by blood volume expansion experienced in pregnancy.

DIAGNOSIS

Anemia is typically diagnosed on a complete blood count. Apart from reporting the number of red blood cells and the hemoglobin level, the automatic counters also measure the size of the red blood cells by flow cytometry, which is an important tool in distinguishing between the causes of anemia. Examination of a stained blood smear using a microscope can also be helpful, and it is sometimes a necessity in regions of the world where automated analysis is less accessible.

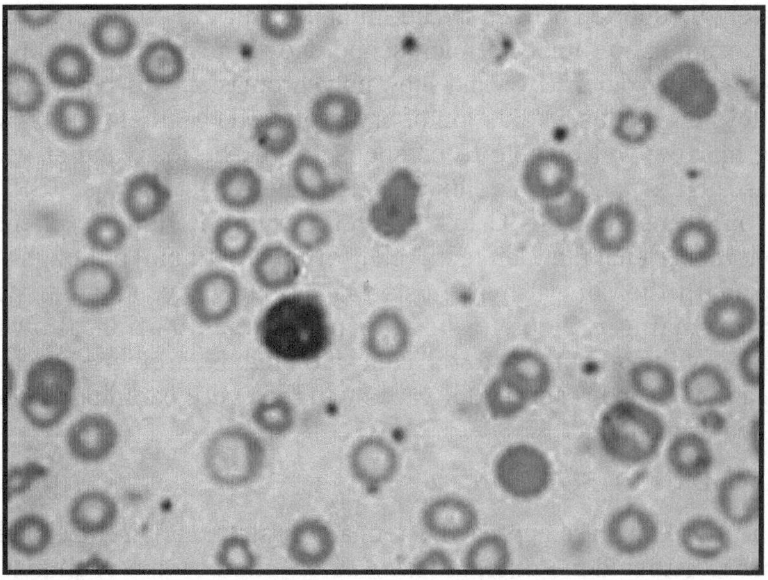

Fig. Peripheral Blood Smear Microscopy of a Patient with Iron-Deficiency Anemia.

In modern counters, four parameters (RBC count, hemoglobin concentration, MCV and RDW) are measured, allowing others (hematocrit, MCH and MCHC) to be calculated, and compared to values adjusted for age and sex. Some counters estimate hematocrit from direct measurements.

WHO's Hemoglobin thresholds used to define anemia (1 g/dL = 0.6206 mmol/L)		
Age or gender group	**Hb threshold (g/dl)**	**Hb threshold (mmol/l)**
Children (0.5–5.0 yrs)	11.0	6.8
Children (5–12 yrs)	11.5	7.1
Teens (12–15 yrs)	12.0	7.4
Women, non-pregnant (>15yrs)	12.0	7.4
Women, pregnant	11.0	6.8
Men (>15yrs)	13.0	8.1

Reticulocyte counts, and the "kinetic" approach to anemia, have become more common than in the past in the large medical centers of the United States and some other wealthy nations, in part because some automatic counters now have the capacity to include reticulocyte counts. A reticulocyte count is a quantitative measure of the bone marrow's production of new red blood cells. The reticulocyte production index is a calculation of the ratio between the level of anemia and the extent to which the reticulocyte count has risen in response. If the degree of anemia is significant, even a "normal" reticulocyte count actually may reflect an inadequate response. If an automated count is not available, a reticulocyte count can be done manually following special staining of the blood film.

In manual examination, activity of the bone marrow can also be gauged qualitatively by subtle changes in the numbers and the morphology of young RBCs by examination under a microscope. Newly formed RBCs are usually slightly larger than older RBCs and show polychromasia. Even where the source of blood loss is obvious, evaluation of erythropoiesis can help assess whether the bone marrow will be able to compensate for the loss, and at what rate. When the cause is not obvious, clinicians use other tests, such as: ESR, ferritin, serum iron, transferrin, RBC folate level, serum vitamin B_{12}, hemoglobin electrophoresis, renal function tests (e.g. serum creatinine) although the tests will depend on the clinical hypothesis that is being investigated. When the diagnosis remains difficult, a bone marrow examination allows direct examination of the precursors to red cells, although is rarely used as is painful, invasive and is hence reserved for cases where severe pathology needs to be determined or excluded.

RED BLOOD CELL SIZE

In the morphological approach, anemia is classified by the size of red blood cells; this is either done automatically or on microscopic examination of a peripheral blood smear. The size is reflected in the mean corpuscular volume (MCV).

If the cells are smaller than normal (under 80 fl), the anemia is said to be microcytic; if they are normal size (80–100 fl), normocytic; and if they are larger than normal (over 100 fl), the anemia is classified as macrocytic. This scheme quickly exposes some of the most common causes of anemia; for instance, a microcytic anemia is often the result of iron deficiency. In clinical workup, the MCV will be one of the first pieces of information available, so even among clinicians who consider the "kinetic" approach more useful philosophically, morphology will remain an important element of classification and diagnosis. Limitations of MCV include cases where the underlying cause is due to a combination of factors - such as iron deficiency (a cause of microcytosis) and vitamin B12 deficiency (a cause of macrocytosis) where the net result can be normocytic cells.

PRODUCTION VS. DESTRUCTION OR LOSS

The "kinetic" approach to anemia yields arguably the most clinically relevant classification of anemia. This classification depends on evaluation of several hematological parameters, particularly the blood reticulocyte (precursor of mature RBCs) count. This then yields the classification of defects by decreased RBC production versus increased RBC destruction and/or loss.

Clinical signs of loss or destruction include abnormal peripheral blood smear with signs of hemolysis; elevated LDH suggesting cell destruction; or clinical signs of bleeding, such as guaiac-positive stool, radiographic findings, or frank bleeding. The following is a simplified schematic of this approach:

- For instance, sickle cell anemia with superimposed iron deficiency; chronic gastric bleeding with B_{12} and folate deficiency; and other instances of anemia with more than one cause.** Confirm by repeating reticulocyte count: ongoing combination of low reticulocyte production index, normal MCV and hemolysis or loss may be seen in bone marrow failure or anemia of chronic disease, with superimposed or related hemolysis or blood loss. Here is a schematic representation of how to consider anemia with MCV as the starting point:

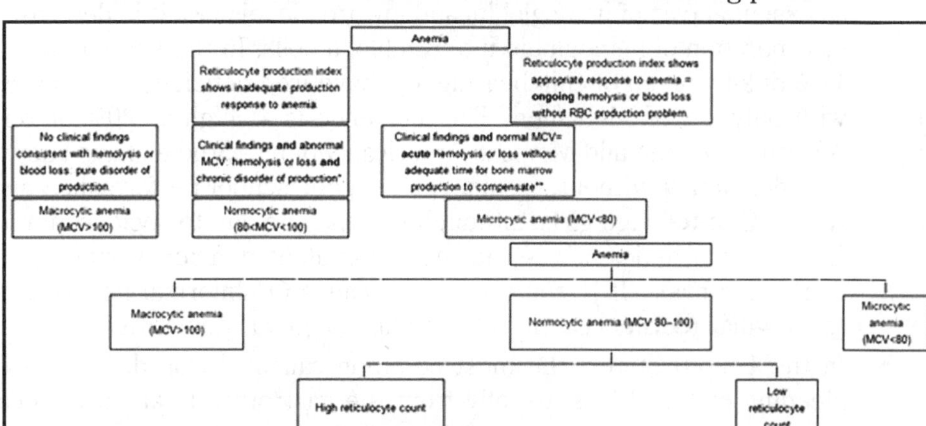

Other characteristics visible on the peripheral smear may provide valuable clues about a more specific diagnosis; for example, abnormal white blood cells may point to a cause in the bone marrow.

Microcytic

Microcytic anemia is primarily a result of hemoglobin synthesis failure/ insufficiency, which could be caused by several etiologies:

- Heme synthesis defect
  - Iron deficiency anemia (microcytosis is not always present)
  - Anemia of chronic disease (more commonly presenting as normocytic anemia)
- Globin synthesis defect
  - Alpha-, and beta-thalassemia
  - HbE syndrome
  - HbC syndrome
  - Various other unstable hemoglobin diseases
- Sideroblastic defect
  - Hereditary sideroblastic anemia
  - Acquired sideroblastic anemia, including lead toxicity
  - Reversible sideroblastic anemia

Iron deficiency anemia is the most common type of anemia overall and it has many causes. RBCs often appear hypochromic (paler than usual) and microcytic (smaller than usual) when viewed with a microscope.

- Iron deficiency anemia is due to insufficient dietary intake or absorption of iron to meet the body's needs. Infants, toddlers, and pregnant women have higher than average needs. Increased iron intake is also needed to offset blood losses due to digestive tract issues, frequent blood donations, or heavy menstrual periods. Iron is an essential part of hemoglobin, and low iron levels result in decreased incorporation of hemoglobin into red blood cells. In the United States, 12% of all women of childbearing age have iron deficiency, compared with only 2% of adult men. The incidence is as high as 20% among African American and Mexican American women. Studies have shown iron deficiency without anemia causes poor school performance and lower IQ in teenage girls, although this may be due to socioeconomic factors. Iron deficiency is the most prevalent deficiency state on a worldwide basis. It is sometimes the cause of abnormal fissuring of the angular (corner) sections of the lips (angular stomatitis).
- In the United States, the most common cause of iron deficiency is bleeding or blood loss, usually from the gastrointestinal tract. Fecal

occult blood testing, upper endoscopy and lower endoscopy should be performed to identify bleeding lesions. In older men and women, the chances are higher that bleeding from the gastrointestinal tract could be due to colon polyps or colorectal cancer.

- Worldwide, the most common cause of iron deficiency anemia is parasitic infestation (hookworms, amebiasis, schistosomiasis and whipworms).

The Mentzer index (mean cell volume divided by the RBC count) predicts whether microcytic anemia may be due to iron deficiency or thallasemia, although it requires confirmation.

Macrocytic

- Megaloblastic anemia, the most common cause of macrocytic anemia, is due to a deficiency of either vitamin B_{12}, folic acid, or both. Deficiency in folate and/or vitamin B_{12} can be due either to inadequate intake or insufficient absorption. Folate deficiency normally does not produce neurological symptoms, while B_{12} deficiency does.
  - Pernicious anemia is caused by a lack of intrinsic factor, which is required to absorb vitamin B_{12} from food. A lack of intrinsic factor may arise from an autoimmune condition targeting the parietal cells (atrophic gastritis) that produce intrinsic factor or against intrinsic factor itself. These lead to poor absorption of vitamin B_{12}.
  - Macrocytic anemia can also be caused by removal of the functional portion of the stomach, such as during gastric bypass surgery, leading to reduced vitamin B_{12}/folate absorption. Therefore, one must always be aware of anemia following this procedure.
- Hypothyroidism
- Alcoholism commonly causes a macrocytosis, although not specifically anemia. Other types of liver disease can also cause macrocytosis.
- Drugs such as Methotrexate, zidovudine, and other substances may inhibit DNA replication such as heavy metals (e.g. Lead)

Macrocytic anemia can be further divided into “megaloblastic anemia” or “nonmegaloblastic macrocytic anemia”. The cause of megaloblastic anemia is primarily a failure of DNA synthesis with preserved RNA synthesis, which results in restricted cell division of the progenitor cells. The megaloblastic anemias often present with neutrophil hypersegmentation (six to 10 lobes). The nonmegaloblastic macrocytic anemias have different etiologies (i.e. unimpaired DNA globin synthesis,) which occur, for example, in alcoholism. In addition to the nonspecific symptoms of anemia, specific features of vitamin B_{12} deficiency include peripheral neuropathy and subacute combined degeneration of the cord with resulting balance difficulties from posterior column spinal cord pathology. Other features may include a smooth, red tongue and glossitis. The treatment for vitamin B_{12}-deficient anemia was first devised by William Murphy, who bled dogs to make them anemic, and then fed them various substances to see what (if anything) would make them healthy again. He discovered that ingesting large amounts of liver seemed to cure the disease. George Minot and George Whipple then set about to isolate the curative substance chemically and ultimately were able to isolate the vitamin B_{12} from the liver. All three shared the 1934 Nobel Prize in Medicine.

Normocytic

Normocytic anemia occurs when the overall hemoglobin levels are decreased, but the red blood cell size (mean corpuscular volume) remains normal. Causes include:

- Acute blood loss
- Anemia of chronic disease
- Aplastic anemia (bone marrow failure)
- Hemolytic anemia

Dimorphic

A dimorphic appearance on a peripheral blood smear occurs when there are two simultaneous populations of red blood cells, typically of different size and hemoglobin content (this last feature affecting the color of the red blood cell on a stained peripheral blood smear). For example, a person recently transfused for iron deficiency would have small, pale, iron deficient red blood cells (RBCs) and the donor RBCs of normal size and color. Similarly, a person transfused for severe folate or vitamin B12 deficiency would have two cell populations, but, in this case, the patient's RBCs would be larger and paler than the donor's RBCs.

A person with sideroblastic anemia (a defect in heme synthesis, commonly caused by alcoholism, but also drugs/toxins, nutritional deficiencies, a few acquired and rare congenital diseases) can have a dimorphic smear from the sideroblastic anemia alone. Evidence for multiple causes appears with an elevated RBC distribution width (RDW), indicating a wider-than-normal range of red cell sizes, also seen in common nutritional anemia.

Heinz Body Anemia

Heinz bodies form in the cytoplasm of RBCs and appear as small dark dots under the microscope. Heinz body anemia has many causes, and some forms can be drug-induced. It is triggered in cats by eating onions or acetaminophen (paracetamol). It can be triggered in dogs by ingesting onions or zinc, and in horses by ingesting dry red maple leaves.

Hyperanemia

Hyperanemia is a severe form of anemia, in which the hematocrit is below 10%.

Refractory Anemia

Refractory anemia, an anemia which does not respond to treatment, is often seen secondary to myelodysplastic syndromes. Iron deficiency anemia may also be refractory as a clinical manifestation of gastrointestinal problems which disrupt iron absorption or cause occult bleeding.

TREATMENTS

Treatments for anemia depend on cause and severity. Vitamin supplements given orally (folic acid or vitamin B_{12}) or intramuscularly (vitamin B_{12}) will replace specific deficiencies.

ORAL IRON

Nutritional iron deficiency is common in developing nations. An estimated two-thirds of children and of women of childbearing age in most developing nations are estimated to suffer from iron deficiency; one-third of them have the more severe form of the disorder, anemia. Iron deficiency from nutritional causes is rare in men and postmenopausal women.

The diagnosis of iron deficiency mandates a search for potential sources of loss, such as gastrointestinal bleeding from ulcers or colon cancer. Mild to moderate iron-deficiency anemia is treated by oral iron supplementation with ferrous sulfate, ferrous fumarate, or ferrous gluconate. When taking iron supplements, stomach upset and/or darkening of the feces are commonly experienced. The stomach upset can be alleviated by taking the iron with food; however, this decreases the amount of iron absorbed. Vitamin C aids in the body's ability to absorb iron, so taking oral iron supplements with orange juice is of benefit.

In anemias of chronic disease, associated with chemotherapy, or associated with renal disease, some clinicians prescribe recombinant erythropoietin or epoetin alfa, to stimulate RBC production, although since there is also concurrent iron deficiency and inflammation present, parenteral iron is advised to be taken concurrently.

INJECTABLE IRON

In cases where oral iron has either proven ineffective, would be too slow (for example, pre-operatively) or where absorption is impeded (for example in cases of inflammation), parenteral iron can be used. The body can absorb up to 6 mg iron daily from the gastrointestinal tract. In many cases the patient has a deficit of over 1,000 mg of iron which would require several months to replace.

This can be given concurrently with erythropoietin to ensure sufficient iron for increased rates of erythropoiesis.

BLOOD TRANSFUSIONS

Blood transfusions in those without symptoms is not recommended until the hemoglobin is below 60 to 80 g/L (6 to 8 g/dL). In those with coronary artery disease who are not actively bleeding transfusions are only recommended when the hemoglobin is below 70 to 80g/L (7 to 8 g/dL). Transfusing earlier does not improve survival.

Transfusions otherwise should only be undertaken in cases of cardiovascular instability.

ERYTHROPOIESIS-STIMULATING AGENT

The motive for the administration of an erythropoiesis-stimulating agent (ESA) is to maintain hemoglobin at the lowest level that both minimizes transfusions and meets the individual persons needs. They should not be used for mild or moderate anemia. They are not recommended in people with chronic kidney disease unless hemoglobin levels are less than 10 g/dL or they have symptoms of anemia. Their use should be along with parenteral iron.

HYPERBARIC OXYGEN

Treatment of exceptional blood loss (anemia) is recognized as an indication for hyperbaric oxygen (HBO) by the Undersea and Hyperbaric Medical Society. The use of HBO is indicated when oxygen delivery to tissue is not sufficient in patients who cannot be given blood transfusions for medical or religious reasons. HBO may be used for medical reasons when threat of blood product incompatibility or concern for transmissible disease are factors. The beliefs of some religions (ex: Jehovah's Witnesses) may require they use the HBO method. A 2005 review of the use of HBO in severe anemia found all publications reported positive results.

EPIDEMIOLOGY

A moderate degree of iron-deficiency anemia affected approximately 610 million people worldwide or 8.8% of the population. It is slightly more common in female (9.9%) than males (7.8%). Mild iron deficiency anemia affects another 375 million.

ASPECTS OF ANEMIAS

ACUTE POSTHEMORRHAGIC ANEMIA

Acute posthemorrhagic anemia or acute blood loss anemia is a condition in which a person quickly loses a large volume of circulating hemoglobin. Acute blood loss is usually associated with an incident of trauma or a severe injury resulting in a large loss of blood. It can also occur during or after a surgical procedure.

ANEMIA OF CHRONIC DISEASE

Anemia of chronic disease, or anemia of chronic inflammation, is a form of anemia seen in chronic infection, chronic immune activation, and malignancy. These conditions all produce massive elevation of Interleukin-6, which stimulates hepcidin production and release from the liver, which in turn reduces the iron carrier protein ferroportin so that access of iron to the circulation is reduced. Other mechanisms may also play a role, such as reduced erythropoiesis. Anemia of chronic inflammation is the preferred term since not all chronic diseases are associated with this form of anemia.

Pathophysiology

Anemia is considered when RBCs count:

- < 4.5 million in males
- < 3.9 million in females

Or Hemoglobin (Hb) content:

- < 13.5 gm % in males
- < 11.5 gm % in females

In response to inflammatory cytokines, increasingly IL-6, the liver produces increased amounts of hepcidin. Hepcidin in turn causes increased internalisation of ferroportin molecules on cell membranes which prevents release from iron stores. Inflammatory cytokines also appear to affect other important elements of iron metabolism, including decreasing ferroportin expression, and probably directly blunting erythropoiesis by decreasing the ability of the bone marrow to respond to erythropoietin.

Before the recent discovery of hepcidin and its function in iron metabolism, anemia of chronic disease was seen as the result of a complex web of inflammatory changes. Over the last few years, however, many investigators have come to feel that hepcidin is the central actor in producing anemia of chronic inflammation. Hepcidin offers an attractive Occam's Razor (parsimonious) explanation for the condition, and more recent descriptions of human iron metabolism and hepcidin function reflect this view.

In addition to effects of iron sequestration, inflammatory cytokines promote the production of white blood cells. Bone marrow produces both white blood cells and red blood cells from the same precursor stem cells. Therefore, the upregulation of white blood cells causes fewer stem cells to differentiate into red blood cells. This effect may be an important additional cause for the decreased erythropoiesis and red blood cell production seen in anemia of inflammation, even when erythropoietin levels are normal, and even aside from the effects of hepcidin. Nonetheless, there are other mechanisms that also contribute to the lowering of hemoglobin levels during inflammation: (i) Inflammatory cytokines suppress the proliferation of erythroid precursors in the bone marrow; (ii) inflammatory cytokines inhibit the release of erythropoietin (EPO) from the kidney; and (iii) the survival of circulating red cells is shortened.

In the short term, the overall effect of these changes is likely positive: it allows the body to keep more iron away from bacterial pathogens in the body, while producing more immune cells to fight off infection. Almost all bacteria depend on iron to live and multiply. However, if inflammation continues, the effect of locking up iron stores is to reduce the ability of the bone marrow to produce red blood cells. These cells require iron for their massive amounts of hemoglobin which allow them to transport oxygen.

Because anemia of chronic disease can be the result of non-infective causes of inflammation, future research is likely to investigate whether hepcidin antagonists might be able to treat this problem.

Anemia of chronic disease may also be due to neoplastic disorders and non-infectious inflammatory diseases. Neoplastic disorders include Hodgkin's disease lung and breast carcinoma and non-infectious inflammatory diseases include rheumatoid arthritis and systemic lupus erythematosus.

Anemia of chronic disease as it is now understood is to at least some degree separate from the anemia seen in renal failure in which anemia results from poor production of erythropoietin, or the anemia caused by some drugs (like AZT, used to treat HIV infection) that have the side effect of inhibiting erythropoiesis. In other words, not all anemia seen in people with chronic disease should be diagnosed as anemia of chronic disease. On the other hand, both of these examples show the complexity of this diagnosis: HIV infection itself can produce anemia of chronic disease, and renal failure can lead to inflammatory changes that also can produce anemia of chronic disease.

Survival Advantage

Limiting some microbes' access to iron can reduce their virulence, thereby potentially reducing the severity of infection. Blood transfusion to patients with anemia of chronic disease is associated with a higher mortality, supporting the concept.

Severity

Anemia of chronic disease is usually mild but can be severe. It is usually normocytic, but can be microcytic. The presence of both anemia of chronic disease and dietary iron deficiency in the same patient results in a more severe anemia.

Diagnosis

While no single test is reliable to distinguish iron deficiency anemia from the anemia of chronic inflammation, there are sometimes some suggestive data:

- In anemia of chronic inflammation without iron deficiency, ferritin is normal or high, reflecting the fact that iron is sequestered within cells, and ferritin is being produced as an acute phase reactant. In iron deficiency anemia ferritin is low.
- TIBC is high in iron deficiency, reflecting production of more transferrin to increase iron binding.; TIBC is low or normal in anemia of chronic inflammation.

Treatment

The ideal treatment for anemia of chronic disease is to treat the chronic disease successfully, but this is rarely possible. Parenteral iron is increasingly used for anemia in chronic renal disease and inflammatory bowel disease. Erythropoietin can be helpful, but this is costly and may be dangerous. Erythropoietin is advised either in conjunction with adequate iron replacement which in practice is intravenous, or when IV iron has proved ineffective.

ANEMIA OF PREMATURITY

Anemia of prematurity refers to a form of anemia affecting preterm infants with decreased hematocrit.

Preterm infants are often anemic and typically experience heavy blood losses from frequent laboratory testing in the first few weeks of life. Although their anemia is multifactorial, repeated blood sampling and reduced erythropoiesis with extremely low serum levels of erythropoietin (EPO) are major determining factors. Blood sampling done for laboratory testing can easily remove enough blood to produce anemia. Obladen, Sachsenweger and Stahnke (1987) studied 60 very low birth weight infants during the first 28 days of life. Infants were divided into 3 groups, group 1 (no ventilator support, 24 ml/kg blood loss), group 2(minor ventilated support, 60 ml/kg blood loss), and group 3(ventilated support for respiratory distress syndrome, 67 ml/kg blood loss). Infants were checked for clinical symptoms and laboratory signs of anemia 24 hours before and after the blood transfusion. The study found that groups 2 and 3 who had significant amount of blood loss, showed poor weight gain, pallor and distended abdomen. These reactions are the most frequent symptoms of anemia.

During the first weeks of life, all infants experience a decline in circulating red blood cell (RBC) volume generally expressed as blood hemoglobin concentration (Hb). As anemia develops, there is even more of a significant reduction in the concentration of hemoglobin. Normally this stimulates a significant increased production of erythropoietin (EPO), but this response is diminished in premature infants. Dear, Gill, Newell, Richards and Schwarz (2005) conducted a study to show that there is a weak negative correlation between EPO and Hb.

The researchers recruited 39 preterm infants from 10 days of age or as soon as they could manage without respiratory support. They estimated total EPO and Hb weekly and 2 days after a blood transfusion. The study found that when Hb>10, EPO mean was 20.6 and when Hbd"10, EPO mean was 26.8. As Hb goes down, EPO goes up. While the reason for this decreased response is not fully understood, Strauss (n.d.) states that it results from both physiological factors (e.g., the rapid rate of growth and need for a commensurate increase in RBC mass to accompany the increase in blood volume) and, in sick premature infants, from phlebotomy blood losses.

In premature infants this decline occurs earlier and more pronounced that it does in healthy term infants. Healthy term infants Hb rarely falls below 9 g/dL at an age of approximately 10–12 weeks, while in premature infants, even in those without complicating illnesses, the mean Hb falls to approximately 8g/dL in infants of 1.0-1.5 kg birth weight and to 7g/dL in infants <1.0 kg. Because this postnatal drop in hemoglobin level is universal and is well tolerated in term infants, it is commonly referred to as the "physiologic" anemia of infancy. However, in premature infants the decline in Hb may be associated with abnormal clinical signs severe enough to prompt transfusions.

Treatment

Transfusion

AOP is usually treated by blood transfusion but the indications for this are still unclear. Blood transfusions have the risk of incompatibility and transfusion reactions as well as viral infections. Also, blood transfusions are costly and add to parental anxiety. The best treatment in prevention is minimizing the amount of blood drawn from the infant. It is found that since blood loss attributable to laboratory testing is the primary cause of anemia among preterm infants during the first weeks of life, we quantified blood lost attributable to phlebotomy overdraw, something that might be avoided. A study was done to see when and if overdraw was a problem.

They recorded all of the data that could be of influence such as the test performed, the blood collection container used, the infants location (neonatal intensive care unit (NICU) and intermediate intensive care unit), the infant's weight sampling and the phlebotomist's level of experience, work shift, and clinical role. Infants were classified by weight into 3 groups: <1 kg, 1 to 2 kg, and >2 kg. The volume of blood removed was calculated by subtracting the weight of the empty collection container from that of the container filled with blood. They found that the mean volume of blood drawn for the 578 tests exceeded that requested by the hospital laboratory by 19.0% ± 1.8% per test. The main factors of overdraw was: collection in blood containers without fill-lines, lighter weight infants and critically ill infants being cared for in the NICU.

EPO

Recombinant EPO (r-EPO) may be given to premature infants to stimulate red blood cell production. Brown and Keith (1999) studied two groups of 40 very low birth weight (VLBW) infants to compare the erythropoietic response between two and five times a week dosages of recombinant human erythropoietin (r-EPO) using the same dose. They established that more frequent dosing of the same weekly amount of r-EPO generated a significant and continuous increase in Hb in VLBW infants. The infants that received five dosages had 219,857 mm^3 while infants that received two dosages only had 173,361 mm^3.

However, the response to r-EPO typically takes up to two weeks and the higher dosages lead to higher Hb. Brown and Keith (1999) study also showed responses between two dosage schedules (two times a week and five times a week). Infants were recruited for gestational age—age since conception—d”27 weeks and 28 to 30 weeks and then randomized into the two groups, each totaling 500 U/kg a week. Brown and Keith found that after two weeks of r-EPO administration, Hb counts had increased and leveled off; the infants who received r-EPO five times a week had significantly higher Hb counts. This was present at four weeks for all infants d”30 weeks gestation and at 8 weeks for infants d”27 weeks gestation.

To date, studies of r-EPO use in premature infants have had mixed results. Ohls et al. examined the use of early r-EPO plus iron and found no short-term benefits in two groups of infants (172 infants less than 1000 g and 118 infants 1000–1250 g). All r-EPO treated infants received 400 U/g three times a week until they reached 35 weeks gestational age. The use of r-EPO did not decrease the average number of transfusions in the infants born at less than 1000 g, or the percentage of infants in the 1000 to 1250 group.

A multi-center European trial studied early versus late r-EPO in 219 infants with birth weights between 500 and 999 g. An r-EPO close of 750 U/kg/week was given to infants in both the early (1–9 weeks) and late (4–10 weeks) groups. The two r-EPO groups were compared to a control group who did not receive r-EPO. Infants in all three groups received 3 to 9 mg/kg of enteral iron. These investigators reported a slight decrease in transfusion and donor exposures in the early r-EPO group (1–9 weeks): 13% early, 11% late and 4% control group. It is likely that only a carefully selected subpopulation of infants may benefit from its use. Contrary to what just said, Bain and Blackburn (2004) also state in another study the use of r-EPO does not appear to have a significant effect on reducing the numbers of early transfusions in most infants, but may be useful to reduce numbers of late transfusion in extremely low-birth-weight infants. A British task force to establish transfusion guidelines for neonates and young children and to help try to explain this confusion recently concluded that "the optimal dose, timing, and nutritional support required during EPO treatment has yet to be defined and currently the routine use of EPO in this patient population is not recommended as similar reduction in blood use can probably be achieved with appropriate transfusion protocols."

Transfusion Management

Other strategies involve the reduction of blood loss during phlebotomy. Another treatment used is therapeutic strategies. These strategies are aimed at reducing transfusions have assessed the use of strict blood transfusions guidelines and EPO therapy, but reduction of blood loss is most important. For extremely low birth weight infants, laboratory blood testing using bedside devices offers a unique opportunity to reduce blood transfusions.

This practice has been referred to as point-of-care testing. Use of these kind of devices to measure the most common ordered blood tests could significantly decrease phlebotomy loss and lead to a reduction in the need for blood transfusions among critically ill premature neonates. A study was done by Adams, Benitz, Geaghan, Kumar, Madan and Widness (2005) to test this theory by conducting a retrospective chart review on all inborn infants <1000g admitted to the NICU during two separate years. Conventional bench top laboratory analysis during the first year was done using Radiometer Blood Gas and Electrolyte Analyzer. Bedside blood gas analysis during the second year was performed using a point-of-care analyzer. An estimated blood loss in the two groups was determined based on the number of specific blood tests on individual infants.

The study found that there was an estimated 30% reduction in the total volume of blood removed for the blood tests. This study concluded that there is modern technology that can be used instead of blood transfusions and r-EPO.

CONGENITAL DYSERYTHROPOIETIC ANEMIA

Congenital dyserythropoietic anemia (CDA) is a rare blood disorder, similar to the thalassemias. CDA is one of many types of anemia, characterized by ineffective erythropoiesis, and resulting from a decrease in the number of red blood cells (RBCs) in the body and a less than normal quantity of hemoglobin in the blood. This shortage prevents the blood from carrying an adequate supply of oxygen to the body's tissues, resulting in various symptoms of anemia including: tiredness (fatigue), weakness, pale skin, and other similar complications.

Genetics

CDA may be transmitted by both parents autosomal recessively or dominantly and has over four different subtypes, but CDA Type I, CDA Type II, CDA Type III, and CDA Type IV are the most common. CDA type II (CDA II) is the most frequent type of congenital dyserythropoietic anemias. More than 200 cases have been described, but with the exception of a report by the International CDA II Registry, these reports include only small numbers of cases and no data on the lifetime evolution of the disease.

Types include:

Type	OMIM	Gene	Locus
CDAN1	224120	*CDAN1 (gene)*	15q15
CDAN2	224100	*SEC23B*	20p11.2
CDAN3	105600	*KIF23*	15q21
CDAN4	613673	*KLF1*	19p13.13-p13.12

Treatment

Patients with CDA typically get frequent blood transfusions, but this can vary depending on the type that is acquired. Patients report going every 2–4 weeks for blood transfusions, and receive somewhere between 1-2 adult units of blood (1-2 child units for children). In addition, they must undertake chelation therapy to survive; either deferoxamine (i.e. Desferal), deferasirox (i.e. Exjade, Asunra), or deferiprone (i.e. Kelfer, Ferriprox, L1) to eliminate the excess iron that accumulates. In a clinical sense, the disorder is very similar to thalassemia major (Beta-thalassemia) and treated in the same fashion. Removal of the spleen and gallbladder are common, and problems with the liver and heart become increasingly important as the individual ages, due to stresses from low hemoglobin counts and high iron content. Hemoglobin levels can run anywhere between 6.0 g/dl and 8.0 g/dl in untransfused patients, and between 9.0 g/dl and 13.0 g/dl in well transfused patients. It should be noted that the amount of blood received by the patient is not as important as their baseline pre-transfusion hemoglobin level. The closer the patient is to the normal range, the better he or she will feel. This is true for ferritin levels and iron levels in the organs as well. It is important for patients to go regularly for transfusions and to chelate daily in order to maximize good health.

Exjade, an oral chelator approved by the FDA in November 2005, has made chelation therapy much easier for patients, who previously could only chelate subcutaneously. Deferiprone, another oral chelator, has also recently been approved by the FDA on October 14, 2011. While normal ferritin levels run anywhere between 24 and 336 ng/ml, hematologists generally do not begin chelation therapy until ferritin levels reach at least 1000 ng/ml. It is more important to check iron levels in the organs through annual MRI scans (T2* for the heart, R2 for the liver), however, than to simply get regular blood tests to check ferritin levels, which only show a trend, and do not reflect actual organ iron content.

Curative Options

Bone marrow transplant and gene therapy are the only known cures for the disorder, but each have their own risks at this point in time. Bone marrow transplantation is the more established cure, as it has been proven to work in human subjects, whereas researchers are still trying to definitively establish this with gene therapy. It generally requires a 10/10 HLA matched donor, however, who is usually a sibling. As most patients do not have this, they must rely on gene therapy research to potentially provide them with a cure. Gene therapy is still experimental and has largely only been tested in animal models until now.

This type of therapy is especially promising, however, as it allows for the autologous transplantation of the patient's own healthy stem cells rather than requiring an outside donor, thereby bypassing any potential for graft vs. host disease (GVHD). Clinical trials on one 18-year-old boy in France in 2007 have suggested that he is cured of Beta-thalassemia three years later with no complications noted so far. A second trial on another patient in France is due to occur sometime by the year 2016. In the United States, the FDA approved clinical trials on Beta thalassemia patients in 2012. The first study, which takes place at Memorial Sloan Kettering Cancer Center in New York, has begun recruiting human subjects with thalassemia major in July 2012. The study is set to end in 2014.

DIAMOND–BLACKFAN ANEMIA

Diamond–Blackfan anemia (DBA), also known as Blackfan-Diamond anemia, inherited pure red cell aplasia and as inherited erythroblastopenia, is a congenital erythroid aplasia that usually presents in infancy. DBA causes low red blood cell counts (anemia), without substantially affecting the other blood components (the platelets and the white blood cells), which are usually normal. This is in contrast to Shwachman–Bodian–Diamond syndrome, in which the bone marrow defect results primarily in neutropenia, and Fanconi anemia, where all cell lines are affected resulting in pancytopenia. A variety of other congenital abnormalities may also occur in DBA.

Clinical Features

Diamond–Blackfan anemia is characterized by normocytic(cite) or macrocytic(cite) anemia (low red blood cell counts) with decreased erythroid progenitors in the bone marrow. This usually develops during the neonatal period. About 47% of affected individuals also have a variety of congenital abnormalities, including craniofacial malformations, thumb or upper limb abnormalities, cardiac defects, urogenital malformations, and cleft palate. Low birth weight and generalized growth delay are sometimes observed. DBA patients have a modest risk of developing leukemia and other malignancies.

Diagnosis

Typically, a diagnosis of DBA is made through a blood count and a bone marrow biopsy. A diagnosis of DBA is made on the basis of anemia, low reticulocyte (immature red blood cells) counts, and diminished erythroid precursors in bone marrow. Features that support a diagnosis of DBA include the presence of congenital abnormalities, macrocytosis, elevated fetal hemoglobin, and elevated adenosine deaminase levels in red blood cells.

Most patients are diagnosed in the first two years of life. However, some mildly affected individuals only receive attention after a more severely affected family member is identified. About 20–25% of DBA patients may be identified with a genetic test for mutations in the RPS19 gene.

History

First noted by Joseph in 1936,[:485] the condition is however named for Diamond and Blackfan, who described congenital hypoplastic anemia in 1938. Responsiveness to corticosteroids was reported in 1951.[:485] In 1961, Diamond and colleagues presented longitudinal data on 30 patients and noted an association with skeletal abnormalities.

In 1997, a region on chromosome 19 was determined to carry a gene mutated in some DBA. In 1999, mutations in the ribosomal protein S19 gene (RPS19) were found to be associated with disease in 42 of 172 DBA patients. In 2001, a second DBA gene was localized to a region of chromosome 8, and further genetic heterogeneity was inferred. Additional genes were subsequently identified.

Genetics

Most pedigrees suggest an autosomal dominant mode of inheritance with incomplete penetrance. Approximately 10–25% of DBA occurs with a family history of disease. With the exception of DBA rarely arising from mutation of transcription factor GATA1,(add citation) DBA arises from abnormal ribosomal protein genes. The disease is characterized by genetic heterogeneity, affecting different ribosomal gene loci:

In 1997, a patient was identified who carried a rare balanced chromosomal translocation involving chromosome 19 and the X chromosome. This suggested that the affected gene might lie in one of the two regions that were disrupted by this cytogenetic anomaly. Linkage analysis in affected families also implicated this region in disease, and led to the cloning of the first DBA gene. About 20–25% of DBA cases are caused by mutations in the ribosome protein S19 (RPS19) gene on chromosome 19 at cytogenetic position 19q13.2. Some previously undiagnosed relatives of DBA patients were found to carry mutations, and also had increased adenosine deaminase levels in their red blood cells, but had no other overt signs of disease.

DBA types					
Name	Chromosome	Genotype	Phenotype	Protein	Disruption(cite)(cite)
DBA1	19q13.2	603474	105650	RPS19	30S to 18S[291](cite)
DBA2	8p23-p22	unknown	606129		
DBA3	10q22-q23	602412	610629	RPS24	30S to 18S[291](cite)
DBA4	15q	180472	612527	RPS17	30S to 18S[291]
DBA5	3q29-qter	180468	612528	RPL35A	32S to 5.8S/28S[291](cite)
DBA6	1p22.1	603634	612561	RPL5	32S to 5.8S/28S[291]
DBA7	1p36.1-p35	604175	612562	RPL11	32S to 5.8S/28S[291]
DBA8	2p25	603658	612563	RPS7	30S to 18S[291]
DBA9	6p	603632	613308	RPS10	30S to 18S
DBA10	12q	603701	613309	RPS26	30S to18S
DBA11	17p13	603704	614900	RPS26	30S to 18S
DBA12	3p24	604174	615550	RPL15	45S to 32S
DBA13	14q	603633	615909	RPS29	
"other"				TSR2,RPS28, GATA1	

A subsequent study of families with no evidence of RPS19 mutations determined that 18 of 38 families showed evidence for involvement of an unknown gene on chromosome 8 at 8p23.3-8p22. The precise genetic defect in these families has not yet been delineated.

Malformations are seen more frequently with DBA6 RPL5 and DBA7 RPL11 mutations.

The genetic abnormalities underpinning the combination of DBA with Treacher Collins syndrome (TCS)/mandibulofacial dysostosis (MFD) phenotypes are heterogeneous, including RPS26 (the known DBA10 gene), TSR2 which encodes a direct binding partner of RPS26, and RPS28. DBA Mutation Database.

Molecular basis

The phenotype of DBA patients suggests a hematological stem cell defect specifically affecting the erythroid progenitor population. Loss of ribosomal function might be predicted to affect translation and protein biosynthesis broadly and impact many tissues.

However, DBA is characterized by dominant inheritance, and arises from partial loss of ribosomal function, so it is possible that erythroid progenitors are more sensitive to this decreased function, while most other tissues are less affected.

Treatment

Corticosteroids can be used to treat anemia in DBA. In a large study of 225 patients, 82% initially responded to this therapy, although many side effects were noted. Some patients remained responsive to steroids, while efficacy waned in others. Blood transfusions can also be used to treat severe anemia in DBA. Periods of remission may occur, during which transfusions and steroid treatments are not required.

Bone marrow transplantation (BMT) can cure hematological aspects of DBA. This option may be considered when patients become transfusion-dependent because frequent transfusions can lead to iron overloading and organ damage. However, adverse events from BMTs may exceed those from iron overloading. An article published on Feb. 10, 2009 reported that an eight-year-old boy with a DBA-like disease has been successfully treated by supplementing his diet with the amino acids leucine and isoleucine. A 2007 study shows the efficacy of a similar treatment on a different patient. Larger studies are being conducted.

INEFFECTIVE ERYTHROPOIESIS

Ineffective erythropoiesis is active erythropoiesis with premature death of red blood cells, a decreased output of RBCs from the bone marrow, and, consequently, anemia.

IMERSLUND–GRÄSBECK SYNDROME

Imerslund–Gräsbeck syndrome, also known as Imerslund–Najman–Gräsbeck syndrome, Imerslund–Gräsbeck disease (IGS or INGS), Imerslund syndrome, congenital cobalamin malabsorption or autosomal recessive megaloblastic anemia (MGA1), is a rare autosomal recessive, familial form of vitamin B_{12} deficiency caused by malfunction of the "Cubam" receptor located in the terminal ileum.

This receptor is composed of two proteins, amnionless (AMN), and cubilin. A defect in either of these protein components can cause this syndrome. This is a rare disease, with a prevalence about 1 in 200,000, and is usually seen in patients of European ancestry.

Vitamin B_{12} is an important vitamin needed for bone marrow functioning, the deficit of which causes decreased marrow output and anemia. Vitamin B_{12} has two forms, one of which, along with folate, is important in DNA synthesis. Vitamin B_{12} is sensitive to acid deformation in the stomach, so a molecule called haptocorrin (R-factor), protects it in the stomach. In the small bowel, a molecule named intrinsic factor (IF), allows vitamin B_{12} to be absorbed in the ileum. IGS is caused by a mutation in the receptors located in the terminal portion of ileum. This is a very rare, and unlikely cause of vitamin B_{12} deficiency but is a cause nonetheless.

Signs and Symptoms

Defined as those seen in any macrocytic, megaloblastic anemia:

- Anemia: causing fatigue, conjuctival pallor, pale complexion, and in some cases, a mild icterus (yellowing of the eye).
- Glossitis ("shiny tongue"): shiny, glossy tongue.
- Cheilosis (stomatitis): Inflammation of the edges of the lips and the oral mucosa.
- Tabes dorsalis ("subacute combined degeneration of the spinal cord"): This involves the posterior section of the spinal cord and therefore involves proprioception (sense of position), touch, sense of vibration and in severe cases the lateral corticospinal tract, causing spastic paralysis of the limbs.
- Peripheral neuropathy: tingling sensation in the arms and legs.
- Pancytopenia: decreased number of blood cells of all lineages (RBCs,leucocytes,platelets), due to decreased bone marrow production.
- Methylmalonyl CoA-emia: defined as blood having and unusually high concentration of methylmalonyl CoA.
- Peripheral findings such as hypersegmented neutrophils and large RBCs on high field view of the blood smears.
- Laboratory findings indicating increased MCV (Mean Corpuscular Volume), decreased Hgb/Hct (indicating anemia), and decreased value of vitamin B_{12} in the blood.
- Proteinuria: protein found in the urine detected by analysis or by dipstick.
- Reversal of all symptoms except neurological symptoms, by IV injection of vitamin B_{12}.
- Schilling test indicating no radioactive vitamin B_{12} in the urine. (This test has dropped out of favor and should not be tried in patients with any form of renal failure).

Genetics

The disease is autosomal recessive, and can therefore skip generations. Mutations in either amnionless (AMN) or cubilin can be the culprit. Due to its autosomal recessive pattern of inheritance, affected individuals (persons possessing a homozygous recessive genotype) need to undergo genetic counselling to identify risk of family members who might be heterozygous genetic carriers. Certain mutations on the CUBN or AMN (genes that encode cubilin and amnionless respectively) have been identified through genetic analysis, and ethnic susceptibility of some of the mutations were indicated from the current research. It has been further suggested that mutations on CUBN were restricted to exon 1-28 which encoded amnionless binding domains (EGF) and IF-Cbl binding region of cubilin,while AMN mutations primarily clustered in intron 8-11 and transmembrane domain in exon 10. Some interesting aspects of particular mutations were also elucidated by the researchers, for example, CUBN mutation c.3890C>T; p.Pro1297Leu,was considered to be a Finnish founder mutation presenting mostly in homozygous status, while an AMN mutation c.208-2A>G, which was thought to account for 15% IGS cases around the world, was postulated as an ancient founder mutation that can trace back to approximately 13,600 years ago.

Pathogenesis

Pathogenesis

Vitamin B_{12}, is an essential water-soluble vitamin found in animal products (such as liver, meat, fish, and dairy products). Vitamin B_{12} is not found in plant sources; a vegetarian diet can be a risk factor for vitamin B_{12} deficiency. Normal daily intake of vitamin B_{12} is 7–30 micro gram, cooking has minimal effect on the structure of this vitamin. The minimal daily adult requirement is 1–3 micro gram, and the human body is able to store at any one time about 2–3 milligram, which is sufficient for at least 2 years of impeccable functioning before the source is depleted. In terms of absorption, no more than 2–3 microgram of vitamin B_{12} can be absorbed on a daily basis, with some 5–10 microgram of the vitamin B_{12} involved in enterohepatic circulation. This is in general a principal characteristic of water-soluble vitamins, in that no matter the oral intake, there is a certain threshold for intestinal absorption hence, low or non-existent chance of intoxication, as opposed to fat-soluble vitamins.

Vitamin B_{12} has a major function in the nuclear replication of the DNA. It is therefore logical that its deficiency causes decrease bone marrow production, one of the most common manifestations of which is decreased red blood cell production or as it is referred to medically, anemia. Vitamin B_{12} however has two major forms in the human body:

- Deoxyadenosyl B_{12} or as it is sometimes referred to Ado B_{12}: Ado B_{12} is essential for acid-base maintenance of the blood, simply because Ado B_{12} is the catalyst that assists the conversion of, Methylmalonyl CoA, into Succinyl CoA. In absence of vitamin B_{12}, levels of Methylmalonyl CoA increase, and this is in fact a great way to distinguish folate deficiency macrocytic anemia, from vitamin B_{12} anemia. The following is the reaction in which Ado B_{12}, plays a pivotal role:

Propionyl CoA ! Methylmalonyl CoA ! Succinyl CoA

- Methyl B12: This form of vitamin B_{12} is essential for conversion of Methy-THF (methyl tetrahydrofolate) into THF, and methyl (CH_3).

The methyl group, is then used to add a carbon, to homocysteine, converting it into Methionine. Methionine is further converted to S-adenosyl methionine, which in turn gives of the extra carbon it received from THF, now to a DNA nucleotide, becoming S-adenosyl homcysteine. S-adenosyl Homocysteine, further loses its "S-adenosyl" attachment, to become homocysteine, and the cycle repeats yet again!

Methyl THF → CH_3 + THF
↓
Homocysteine → Methionine
S-adenosyl ← ↑ ↓ ← S-adenosyl
S-adenosyl Homocysteine ← S-adenosyl Methionine
↓
CH_3
↓
DNA → Methyl-DNA

It is therefore understood that vitamin B_{12} is involved in complex DNA synthesis, along with folate, as well as in acid-base metabolism. To understand the basic pathophysiology of Imerslund-Gräsbeck syndrome, it is imperative to understand the absorption of vitamin B_{12}. The following lists principal events that lead to absorption of vitamin B_{12} along the GI tract:

- Oral cavity: vitamin B_{12} containing food is ingested. Salivary glands produce haptocorrin, which binds vitamin B_{12}, creating a "vitamin B_{12}-Haptocorrin complex". This complex is then ingested via esophageal peristalsis into the stomach.
- Stomach: vitamin B_{12}-Haptocorrin, survives the low pH, highly osmotic environment of the stomach. Parietal cells produce hydrochloric acid (the effect of which Haptocorrin protects vitamin B_{12} from), and also intrinsic factor (IF). Intrinsic factor also has a high binding affinity for vitamin B_{12}, but because that position is already filled by Haptocorrin, free intrinsic factor, and "Haptocorrin-vitamin B_{12}" complex, empty from the stomach into the duodenum.
- duodenum: Pancreatic juice, produced by the pancreas, contains pancreatic proteases that break the haptocorrin, degrading it and freeing the vitamin B_{12}. Once free, vitamin B_{12}, binds with intrinsic factor (IF), to produce an "IF-vitamin B_{12}" complex.
- Ileum: Cubam, a receptor found in the terminal portion of the ileum is a specialized receptor complex. The complex is responsible for the recognition of the "vitamin B_{12}-IF" complex and initiating the endocytosis of the complex, resulting in absorption.

Cubam is composed of two molecules, amnionless (AMN) and cubilin. Cubilin is a multi-ligand protein that contains eight epidermal growth factor

(EGF) repeats and 27 CUB domains, from which the four active domains (CUB5-8) collectively get involved in binding interaction with the IF-Cbl complex. Whereas, amnionless is an apical transmembrane protein which is expressed in both intestine and kidney, and it seems to assist the subcellular localization and endocytosis of the cubilin by binding to its amino-terminal residues. Cubilin specializes in recognition of the "vitamin B_{12}-IF" complex and attaches to it, while amnionless (AMN) is responsible for initiation of the endocytosis of complex and the subsequent absorption of vitamin B_{12}. It is at this point that the pathology of IGS syndrome occurs by preventing the absorption of vitamin B_{12}, and can be caused by a mutation in either the amnionless (AMN) portion or the cubilin portion of the receptor.

Treatment

Since the essential pathology is due to the inability to absorb vitamin B_{12} from the bowels, the solution is therefore injection of IV vitamin B_{12}. Timing is essential, as some of the side effects of vitamin B_{12} deficiency are reversible (such as RBC indices, peripheral RBC smear findings such as hypersegmented neutrophils, or even high levels of methylmalonyl CoA), but some side effects are irreversible as they are of a neurological source (such as tabes dorsalis, and peripheral neuropathy). High suspicion should be exercised when a neonate, or a pediatric patient presents with anemia, proteinuria, sufficient vitamin B_{12} dietary intake, and no signs of pernicious anemia.

Epidemiology

This is a rare disease with prevalence about 1 in 200,000 to 1 in 600,000. Studies showed that mutations in CUBN or AMN clustered particularly in the Scandinavian countries and the Eastern Mediterranean regions. Founder effect, higher clinical awareness to IGS, and frequent consanguineous marriages all play a role in the higher prevalence of IGS among these populations

HYPOCHROMIC ANEMIA

Hypochromic anemia is a generic term for any type of anemia in which the red blood cells (erythrocytes) are paler than normal. (Hypo- refers to less, and chromic means color.) A normal red blood cell will have an area of pallor in the center of it; it is biconcave disk shaped. In hypochromic cells, this area of central pallor is increased.

This decrease in redness is due to a disproportionate reduction of red cell hemoglobin (the pigment that imparts the red color) in proportion to the volume of the cell. In many cases, the red blood cells will also be small (microcytic), leading to substantial overlap with the category of microcytic anemia. The most common causes of this kind of anemia are iron deficiency and thalassemia.

Hypochromic anemia was historically known as chlorosis or green sickness for the distinct skin tinge sometimes present in patients, in addition to more general symptoms such as a lack of energy, shortness of breath, dyspepsia, headaches, a capricious or scanty appetite and amenorrhea.

Historical Understanding

- *Pandar:* Now, the pox upon her green-sickness for me!
- *Bawd:* 'Faith, there's no way to be rid on't but by the way to the pox. Here comes the Lord Lysimachus disguised. Shakespeare (attrib). Pericles Prince of Tyre

In 1554, German physician Johannes Lange described the condition as "peculiar to virgins". He prescribed that sufferers should "live with men and copulate. If they conceive, they will recover." The name "chlorosis" was coined in 1615 by Montpellier professor of medicine Jean Varandal from the word "Chloris" (Greek:χλωριζ) meaning "greenish-yellow," "pale green," "pale," "pallid" or "fresh". Both Lange and Varande claimed Hippocrates as a reference.

In addition to "green sickness", the condition was known as morbus virgineus ("virgin's disease") or febris amatoria ("lover's fever"). Francis Grose' 1811 Dictionary of the Vulgar Tongue defined "green sickness" as: "The disease of maids occasioned by celibacy." In 1681, English physician Thomas Sydenham classified chlorosis as a hysterical disease affecting not only adolescent girls but also "slender and weakly women that seem consumptive." He advocated iron as a treatment: "To the worn out or languid blood it gives a spur or fillip whereby the animal spirits which lay prostrate and sunken under their own weight are raised and excited".

Daniel Turner in 1714 preferred to term chlorosis "the Pale or White Sickness... since in its worst State the Complexion is rarely or ever a true Green, tho' bordering on that Hue". He went on to describe it as "an ill Habit of Body, arising either from Obstructions, particularly of the menstrual Purgation, or from a Congestion of crude Humours in the Viscera, vitiating the Ferments of the Bowels, especially those of Concoction, and placing therein a depraved Appetite of Things directly preternatural, as Chalk, Cinders, Earth, Sand, &c". One of his case studies was that of an 11-year-old girl who was found, on investigation, to have been eating large quantities of coal.

Chlorosis is briefly mentioned in Casanova's Histoire de ma vie: "I do not know, but we have some physicians who say that chlorosis in girls is the result of that pleasure onanism indulged in to excess". In 1841, the Bohemian doctor and pharmacist Albert Popper published a treatment for Chlorosis containing Vitriolum martis (sulfuric acid and iron) and Sal tartari (potassium carbonate) in Österreichische medicinische Wochenschrift which was republished and refined in the following years.

In 1845, the French writer Auguste Saint-Arroman gave a recipe for a treatment by medicinal chocolate that included iron filings in his De L'action du café, du thé et du chocolat sur la santé, et de leur influence sur l'intelligence et le moral de l'homme and in 1872, French physician Armand Trousseau also advocated treatment with iron, although he still classified chlorosis as a "nervous disease".

In 1887, physician Sir Andrew Clark of London Hospital proposed a physiological cause for chlorosis, tying its onset to the demands placed on the bodies of adolescent girls by growth and menarche. In 1891, Frank Wedekind's play Spring Awakening referenced the disease. In 1895, University of Edinburgh pathologist Ralph Stockman built upon experiments demonstrating that inorganic iron contributed to hemoglobin synthesis to show that chlorosis could be explained by a deficiency in iron brought on by loss of menstrual blood and an inadequate diet. Despite the work of Stockman and the effectiveness of iron in treating the symptoms of chlorosis, debate about its cause continued into the 1930s. A character in T. C. Boyle's The Road to Wellville suffers from chlorosis, and the narrator describes her green skin and black lips. In 1936, Arthur J. Patek and Clark W. Heath of Harvard Medical School concluded that chlorosis was identical to hypochromic anemia.

Acquired Forms

Hypochromic anemia may be caused by vitamin B6 deficiency from a low iron intake, diminished iron absorption, or excessive iron loss. It can also be caused by infections (e.g. hookworms) or other diseases, therapeutic drugs, and lead poisoning. One acquired form of anemia is also known as Faber's syndrome. It may also occur from severe stomach or intestinal bleeding caused by ulcers or medications such as aspirin or bleeding from hemorrhoids.

Hereditary Forms

It can also occur in certain forms of congenital developmental disorders, like Benjamin syndrome. It is also caused by thalassemia.

CONGENITAL DYSERYTHROPOIETIC ANEMIA TYPE I

Congenital dyserythropoietic anemia type I (CDA I) is a disorder of blood cell production, particularly of the production of erythroblasts, which are the precursors of the red blood cells (RBCs).

Genetics

CDA type I is transmitted by both parents autosomal recessively and usually results from mutations in the CDAN1 gene. Little is known about the function of this gene, and it is unclear how mutations cause the characteristic features of CDA type I. Some people with this condition do not have identified mutations in the CDAN1 gene, leading researchers to believe that mutations in at least one other gene can also cause this form of the disorder.

Version	Type	OMIM	Gene	Locus
Ia	CDAN1A	224120	*CDAN1*	15q15
Ib	CDAN1B	615631	*C15ORF41*	15q14

Description

CDA type I is characterized by moderate to severe anemia. It is usually diagnosed in childhood or adolescence, although in some cases, the condition can be detected before birth. Many affected individuals have yellowing of the skin and eyes (jaundice) and an enlarged liver and spleen (hepatosplenomegaly). This condition also causes the body to absorb too much iron, which builds up and can damage tissues and organs. In particular, iron overload can lead to an abnormal heart rhythm (arrhythmia), congestive heart failure, diabetes, and chronic liver disease (cirrhosis). Rarely, people with CDA type I are born with skeletal abnormalities, most often involving the fingers and/or toes.

Treatment

Treatment consists of frequent blood transfusions and chelation therapy. Potential cures include bone marrow transplantation and gene therapy.

CONGENITAL DYSERYTHROPOIETIC ANEMIA TYPE II

Congenital dyserythropoietic anemia type II (CDA II), or HEMPAS is a rare genetic anemia in humans characterized by hereditary erythroblastic multinuclearity with positive acidified serum lysis test.

Genetics

CDA type II is caused by mutations in the SEC23B gene. This gene provides instructions for making a protein that is involved in the transport of other proteins within cells. During the development of red blood cells, this protein may help ensure that proteins are transported to the areas where they are needed. Researchers are working to determine how mutations in the SEC23B gene lead to the signs and symptoms of CDA type II.

Analyses of CDA II erythrocyte membranes showed that the band 3 glycoprotein is underglycosylated. An aberrant glycosylation pattern is seen in the polylactosamine carbohydrates which are normally attached to the band 3 and band 4.5 glycoproteins. The polylactosamines are, however, accumulated in the form of glycolipids. Therefore a genetic factor in CDA II appears to block the glycosylation of protein acceptors and shift these carbohydrates to the lipid acceptors. Structural analysis of CDA II band 3 carbohydrates identified truncated hybrid-type oligosaccharides and suggests that the Golgi glycosylation enzyme(s), alpha-mannosidase II or N-acetylglycosaminyltransferase II is defective in CDA II.

Type	OMIM	Gene	Locus
CDAN2	224100	SEC23B	20p11.2

Description

The anemia associated with CDA type II can range from mild to severe, and most affected individuals have jaundice, hepatosplenomegaly, and the formation of hard deposits in the gallbladder called bilirubin gallstones. This form of the disorder is usually diagnosed in adolescence or early adulthood. An abnormal buildup of iron typically occurs after age 20, leading to complications including heart disease, diabetes, and cirrhosis.

Treatment

Treatment consists of frequent blood transfusions and chelation therapy. Potential cures include bone marrow transplantation and gene therapy.

CONGENITAL DYSERYTHROPOIETIC ANEMIA TYPE III

Congenital dyserythropoietic anemia type III (CDA III) is a rare autosomal dominant disorder characterized by macrocytic anemia, bone marrow erythroid hyperplasia and giant multlnucleate erythroblasts. New evidence suggests that this may be passed on recessively as well.

Genetics

CDA type III is transmitted autosomal dominantly. The genetic cause of CDA type III is known to be a problem with the KIF23 gene, located on the long arm of chromosome 15 at a position designated 15q22.

Type	OMIM	Gene	Locus
CDAN3	105600	KIF23	15q21

Description

The signs and symptoms of CDA type III tend to be milder than those of the other types. Most affected individuals do not have hepatosplenomegaly, and iron does not build up in tissues and organs. In adulthood, abnormalities of a specialized tissue at the back of the eye (the retina) can cause vision impairment. Some people with CDA type III also have a blood disorder known as monoclonal gammopathy, which can lead to a cancer of white blood cells (multiple myeloma).

Treatment

Treatment consists of frequent blood transfusions and chelation therapy. Potential cures include bone marrow transplantation and gene therapy.

CONGENITAL DYSERYTHROPOIETIC ANEMIA TYPE IV

Congenital dyserythropoietic anemia type IV (CDA IV) has been described with typical morphologic features of CDA II but a negative acidified-serum test.

Genetics

Congenital dyserythropoietic anemia type IV is an autosomal dominant inherited red blood cell disorder characterized by ineffective erythropoiesis and hemolysis resulting in anemia. Circulating erythroblasts and erythroblasts in the bone marrow show various morphologic abnormalities. Affected individuals with CDAN4 also have increased levels of fetal hemoglobin.

Type	OMIM	Gene	Locus
CDAN4	613673	KLF1	19p13.13-p13.12

Description

CDA type IV is characterized by mild to moderate splenegomegaly. Hemoglobin is very low and patients are transfusion dependent. MCV is normal or mildly elevated. Erythropoiesis is normoblastic or mildly to moderately megaloblastic. Nonspecific erythroblast dysplasia is present.

Treatment

Treatment consists of frequent blood transfusions and chelation therapy. Potential cures include bone marrow transplantation and gene therapy.

MACROCYTIC ANEMIA

The term macrocytic is from Greek words meaning “large cell”. A macrocytic class of anemia is an anemia (defined as blood with an insufficient concentration of hemoglobin) in which the red blood cells (erythrocytes) are larger than their normal volume. The normal erythrocyte volume in humans is about 80 to 100 femtoliters (fL= $10^{"15}$ L). In metric terms the size is given in equivalent cubic micrometers (1 ìm^3 = 1 fL). The condition of having erythrocytes which (on average) are too large, is called macrocytosis. In contrast, in microcytic anemia, the erythrocytes are smaller than normal.

In a macrocytic anemia the larger red cells are always associated with insufficient numbers of cells and often also insufficient hemoglobin content per cell. Both of these factors work to the opposite effect of larger cell size, to finally result in a total blood hemoglobin concentration that is less than normal (i.e., anemia).

Macrocytic anemia is not a disease in the sense of having a single pathology, but is rather a condition. As such, it is the class name for a set of pathologies that all produce somewhat the same red blood cell abnormality. Many specific pathologies are known which result in macrocytic-type anemias. Some of these produce slightly different sets of appearances in blood cells that are detectable from red and white cell morphology, and others are only detectable with chemical testing.

Types of Macrocytic Anemias

Megaloblastic Anemias (DNA Replication Disorders)

Especially common causes of macrocytic anemias are the so-called megaloblastic anemias, in which cells are larger because they cannot produce DNA quickly enough to divide at the right time as they grow, and thus grow too large before division. Causes of the DNA synthetic problem range from lack of certain vitamins needed to produce DNA (notably folate and B12), to poisons or inhibitors of DNA replication, such as some kinds of antiviral drugs and chemotherapeutic agents. Classically these megaloblastic types of anemias are associated also with more specific features, such as megaloblasts in the bone marrow, the presence of ovalocytes in the (peripheral) blood smear, and the pathognomonic presence of hypersegmented neutrophils.

Red Cell Membrane Disorders Producing Codocytes

Other disorders which cause macrocytosis without DNA replication problems (i.e., non-megaloblastic macrocytic anemias), are disorders associated with increased red cell membrane surface area, such as pathologies of the liver and spleen which produce codocytes or "target cells" which have a central collection of hemoglobin surrounded by a pallor (a thin area) then followed by a thicker collection of hemoglobin at the rim of the cell.

Alcohol

Round macrocytes which are not codocytes are produced in chronic alcoholism (which produces a mild macrocytosis even in the absence of vitamin deficiency), apparently as a direct toxic effect of alcohol specifically on the bone marrow.

Association with Rapid Red Cell Turnover and Reticulocytosis

Mild macrocytosis is a common finding associated with rapid blood restoration or production, since in general, "fresh" or newly produced red cells (reticulocytes) are larger than the mean (average) size, due to slow shrinkage of normal cells over a normal red cell circulating lifetime. Thus, chronic obstructive pulmonary disease (COPD), in which red cells are rapidly produced in response to low oxygen levels in the blood, often produces mild macrocytosis.

Also, rapid blood replacement from the marrow after a traumatic blood loss, or rapid red blood cell turnover from rapid hemolysis (G6PD deficiency), also often produces mild macrocytosis in the associated anemia.

NON-SIDEROPENIC HYPOCHROMIC ANAEMIA

- Red cell indices and blood film appearances suggest iron deficiency, although peripheral blood changes are not usually as marked as in moderate or severe iron deficiency.
- Erythropoiesis is abnormal because of ineffective iron utilisation with poor haemoglobinisation of red cell precursors and
- Bone marrow iron stores are normal or increased and sideroblasts may be frequent and abnormal.

Causes

- Secondary anaemias
  - Chronic infection/inflammation
  - Malignancy
- Thalassaemia
- Sideroblastic anaemia

"Differentiation from Iron Deficiency"

The serum iron and total iron-binding capacity (transferrin) are helpful but not diagnostic; it is quiet possible to have co-existing ineffective iron utilisation and iron deficiency, as determined by bone marrow iron status,e.g. in rheumatoid arthritis.

NORMOCHROMIC ANEMIA

Normochromic is a form of anemia in which the concentration of hemoglobin in the red blood cells is within the standard range. However, there are insufficient numbers of red blood cells. This includes:

- Aplastic,
- Posthemorrhagic, and
- Hemolytic anemias and anemia of chronic disease.

NORMOCYTIC ANEMIA

Normocytic anemia is a common issue that occurs for men and women typically over 85 years old. Its prevalence increases with age, reaching 44 percent in men older than 85 years. A normocytic anemia is defined as an anemia with a mean corpuscular volume (MCV) of 80-100 which is the normal range. However, the hematocrit and hemoglobin is decreased.

Causes

The issue is thought of as representing any of the following:

- A decrease production of normal-sized red blood cells (e.g., anemia of chronic disease, aplastic anemia);

- An increased production of HbS as seen in sickle cell disease (not sickle cell trait);
- An increased destruction or loss of red blood cells (e.g., hemolysis, posthemorrhagic anemia);
- An uncompensated increase in plasma volume (e.g., pregnancy, fluid overload);
- A B2 (riboflavin) deficiency
- A B6 (pyridoxine) deficiency
- Or a mixture of conditions producing microcytic and macrocytic anemia.

SPHEROCYTOSIS

Spherocytosis is an auto-hemolytic anemia (a disease of the blood) characterized by the production of spherocytes - red blood cells (RBCs), or erythrocytes, that are sphere-shaped, rather than bi-concave disk shaped. Spherocytes are found in hereditary spherocytosis and autoimmune hemolytic anemia. It almost always refers to hereditary spherocytosis.

This is caused by a molecular defect in one or more of the proteins of the red blood cell cytoskeleton, including, spectrin, ankyrin, Band 3, or Protein 4.2. Because the cell skeleton has a defect, the blood cell contracts to its most surface-tension efficient and least flexible configuration, a sphere. Though the spherocytes have a smaller surface area through which oxygen and carbon dioxide can be exchanged, they in themselves perform adequately to maintain healthy oxygen supplies. However, they have a high osmotic fragility—when placed into water, they are more likely to burst than normal red blood cells. These cells are more prone to physical degradation.

A1C tests are also compromised by the disorder and must not be used to measure time-based blood glucose levels. Instead, diabetes related glucose tests are limited to fructosamine, which measures the amount of glucose in the blood over the past three weeks on average. This is a marked contrast to A1C tests that measure blood glucose over an average span of three months.They are most commonly found in immunologically-mediated hemolytic anemias and in hereditary spherocytosis, but the former would have a positive direct Coombs test and the latter would not.

The misshapen but otherwise healthy red blood cells are mistaken by the spleen for old or damaged red blood cells and it thus constantly breaks them down, causing a cycle whereby the body destroys its own blood supply (auto-hemolysis). The term "non hereditary spherocytosis" is occasionally used, albeit rarely.

5

White Blood Cell

White blood cells (WBCs), also called leukocytes or leucocytes, are the cells of the immune system that are involved in protecting the body against both infectious disease and foreign invaders. All leukocytes are produced and derived from a multipotent cell in the bone marrow known as a hematopoietic stem cell. Leukocytes are found throughout the body, including the blood and lymphatic system.

Five different and diverse types of leukocytes exist. These types are distinguished by their physical and functional characteristics. Monocytes and neutrophils are phagocytic.

The number of leukocytes in the blood is often an indicator of disease, and thus the WBC count is an important subset of the complete blood count. The normal white cell count is usually between 4 and 11 × 10^9/L. In the US this is usually expressed as 4,000–11,000 white blood cells per microliter of blood. They make up approximately 1% of the total blood volume in a healthy adult. An increase in the number of leukocytes over the upper limits is called leukocytosis, and a decrease below the lower limit is called leukopenia.

The name "white blood cell" derives from the physical appearance of a blood sample after centrifugation. White cells are found in the buff, a thin, typically white layer of nucleated cells between the sedimented red blood cells and the blood plasma. The scientific term leukocyte directly reflects its description. It is derived from the Greek word leuko- meaning "white" and kytos meaning "hollow vessel", with -cyte translated as "cell" in modern use. The buffy coat may sometimes be green if there are large amounts of neutrophils in the sample, due to the heme-containing enzyme myeloperoxidase that they produce.

TYPES

All white blood cells are nucleated but are otherwise distinct in form and function. White cells are best classified into two major lineages: the myeloid leukocytes and the lymphocytes. The white cells of the myeloid lineage include neutrophils, monocytes, eosinophils and basophils. Lymphocytes include T cells, B cells and natural killer cells.

NEUTROPHIL GRANULOCYTE

Neutrophil granulocytes (also known as neutrophils) are the most abundant (40% to 75%) type of white blood cells in most mammals and form an essential part of the innate immune system. Functionality varies in different animals. They are formed from stem cells in the bone marrow. They are short-lived and highly motile. Neutrophils may be subdivided into segmented neutrophils and banded neutrophils (or bands). They form part of the polymorphonuclear cell family (PMNs) together with basophils and eosinophils.

The name neutrophil derives from staining characteristics on hematoxylin and eosin (H&E) histological or cytological preparations. Whereas basophilic white blood cells stain dark blue and eosinophilic white blood cells stain bright red, neutrophils stain a neutral pink. Normally, neutrophils contain a nucleus divided into 2–5 lobes.

Neutrophils are a type of phagocyte and are normally found in the bloodstream. During the beginning (acute) phase of inflammation, particularly as a result of bacterial infection, environmental exposure, and some cancers, neutrophils are one of the first-responders of inflammatory cells to migrate towards the site of inflammation. They migrate through the blood vessels, then through interstitial tissue, following chemical signals such as Interleukin-8 (IL-8), C5a, fMLP and Leukotriene B4 in a process called chemotaxis. They are the predominant cells in pus, accounting for its whitish/yellowish appearance. Neutrophils are recruited to the site of injury within minutes following trauma, and are the hallmark of acute inflammation.

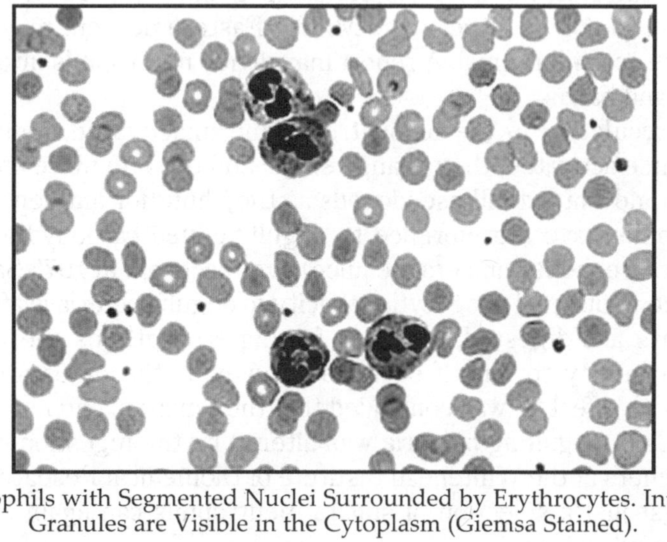

Fig. Neutrophils with Segmented Nuclei Surrounded by Erythrocytes. Intra-Cellular Granules are Visible in the Cytoplasm (Giemsa Stained).

Characteristics

Neutrophil granulocytes have an average diameter of 12-15 micrometers (μm) in peripheral blood smears. When analyzing neutrophils in suspension, neutrophils have an average diameter of 8.85 μm.

With the eosinophil and the basophil, they form the class of polymorphonuclear cells, named for the nucleus' multilobulated shape (as compared to lymphocytes and monocytes, the other types of white cells). The nucleus has a characteristic lobed appearance, the separate lobes connected by chromatin. The nucleolus disappears as the neutrophil matures, which is something that happens in only a few other types of nucleated cells.[:168] In the cytoplasm, the Golgi apparatus is small, mitochondria and ribosomes are sparse, and the rough endoplasmic reticulum is absent.[:170] The cytoplasm also contains about 200 granules, of which a third are azurophilic.[:170] A minor difference is found between the neutrophils from male subjects and female subjects. The cell nucleus of a neutrophil from a female subject shows a small additional X chromosome structure, known as a "neutrophil drumstick".[:174]

Neutrophils will show increasing segmentation (many segments of nucleus) as they mature. A normal neutrophil should have 3-5 segments. Hypersegmentation is not normal, and occurs in some disorders, most notably Vitamin B-12 deficiency. This is noted on a manual review of the blood smear, and is positive when most or all or the neutrophils have 5 or more segments. Neutrophils are the most abundant white blood cells in humans (approximately 10^{11} are produced daily); they account for approximately 50-70% of all white blood cells (leukocytes). The stated normal range for human blood counts varies between laboratories, but a neutrophil count of 2.5–7.5 x 10^9/L is a standard normal range. People of African and Middle Eastern descent may have lower counts, which are still normal. A report may divide neutrophils into segmented neutrophils and bands.

When circulating in the bloodstream and unactivated, neutrophils are spherical. Once activated, they change shape and become more amorphous or amoeba-like and can extend pseudopods as they hunt for antigens.

Neutrophils have a preference to engulf refined carbohydrate (glucose, fructose, sucrose, honey and orange juice) over bacteria. In 1973 Sanchez et al. found that the neutrophil phagocytic capacity to engulf bacteria is affected when simple sugars are digested, and that fasting strengthens the neutrophils' phagocytic capacity to engulf bacteria. However, the digestion of normal starches has no effect. It was concluded that the function, and not the number, of phagocytes in engulfing bacteria was altered by the ingestion of sugars. In 2007 researchers at the Whitehead Institute of Biomedical Research found that even amongst just a selection of sugars, neutrophils can again be picky and engulf preferred sugars.

Life Span

The average lifespan of (non-activated human) neutrophils in the circulation has been reported by different approaches to be between 5 and 90 hours. Upon activation, they marginate (position themselves adjacent to the blood vessel endothelium), and undergo selectin-dependent capture followed by integrin-dependent adhesion in most cases, after which they migrate into tissues, where they survive for 1–2 days.

Neutrophils are much more numerous than the longer-lived monocyte/macrophage phagocytes. A pathogen (disease-causing microorganism or virus) is likely to first encounter a neutrophil. Some experts hypothesize that the short lifetime of neutrophils is an evolutionary adaptation. The short lifetime of neutrophils minimizes propagation of those pathogens that parasitize phagocytes because the more time such parasites spend outside a host cell, the more likely they will be destroyed by some component of the body's defenses. Also, because neutrophil antimicrobial products can also damage host tissues, their short life limits damage to the host during inflammation. Neutrophils will often be phagocytosised by macrophages after digestion of pathogens. PECAM-1 and phosphatidylserine on the cell surface are involved in this process.

Chemotaxis

Neutrophils undergo a process called chemotaxis, which allows them to migrate toward sites of infection or inflammation. Cell surface receptors allow neutrophils to detect chemical gradients of molecules such as interleukin-8 (IL-8), interferon gamma (IFN-gamma), C3a, C5a, and Leukotriene B4, which these cells use to direct the path of their migration.

Neutrophils have a variety of specific receptors, including complement receptors, cytokine receptors for interleukins and interferon gamma (IFN-gamma), receptors for chemokines, receptors to detect and adhere to endothelium, receptors for lectins and proteins, and Fc receptors for opsonin.

Anti-Microbial Function

Being highly motile, neutrophils quickly congregate at a focus of infection, attracted by cytokines expressed by activated endothelium, mast cells, and macrophages. Neutrophils express and release cytokines, which in turn amplify inflammatory reactions by several other cell types.

In addition to recruiting and activating other cells of the immune system, neutrophils play a key role in the front-line defence against invading pathogens. Neutrophils have three methods for directly attacking micro-organisms: phagocytosis (ingestion), degranulation (release of soluble anti-microbials), and generation of neutrophil extracellular traps (NETs).

Phagocytosis

Neutrophils are phagocytes, capable of ingesting microorganisms or particles. For targets to be recognised, they must be coated in opsonins—a process known as antibody opsonization. They can internalize and kill many microbes, each phagocytic event resulting in the formation of a phagosome into which reactive oxygen species and hydrolytic enzymes are secreted. The consumption of oxygen during the generation of reactive oxygen species has been termed the "respiratory burst", although unrelated to respiration or energy production. The respiratory burst involves the activation of the enzyme NADPH oxidase, which produces large quantities of superoxide, a reactive oxygen species. Superoxide decays spontaneously or is broken down via enzymes known as superoxide dismutases (Cu/ZnSOD and MnSOD), to hydrogen peroxide, which is then converted to hypochlorous acid HClO, by the green heme enzyme myeloperoxidase. It is thought that the bactericidal properties of HClO are enough to kill bacteria phagocytosed by the neutrophil, but this may instead be a step necessary for the activation of proteases.

Degranulation

Neutrophils also release an assortment of proteins in three types of granules by a process called degranulation. The contents of these granules have antimicrobial properties, and help combat infection.

Granule type	Protein
azurophilic granules (or "primary granules")	myeloperoxidase, bactericidal/permeability-increasing protein (BPI), defensins, and the serine proteases neutrophil elastase and cathepsin G
specific granules (or "secondary granules")	alkaline phosphatase, lysozyme, NADPH oxidase, collagenase, lactoferrin and cathelicidin
tertiary granules	cathepsin and gelatinase

Neutrophil Extracellular Traps (NETs)

In 2004, Brinkmann and colleagues described a striking observation that activation of neutrophils causes the release of web-like structures of DNA; this represents a third mechanism for killing bacteria. These neutrophil extracellular traps (NETs) comprise a web of fibers composed of chromatin and serine proteases that trap and kill microbes extracellularly. It is suggested that NETs provide a high local concentration of antimicrobial components and bind, disarm, and kill microbes independent of phagocytic uptake. In addition to their possible antimicrobial properties, NETs may serve as a physical barrier that prevents further spread of pathogens.

Trapping of bacteria may be a particularly important role for NETs in sepsis, where NETs are formed within blood vessels. Recently, NETs have been shown to play a role in inflammatory diseases, as NETs could be detected in preeclampsia, a pregnancy-related inflammatory disorder in which neutrophils are known to be activated. In addition, NETs are known to exhibit prothrombotic effects both in vitro and in vivo.

Role in Disease

Low neutrophil counts are termed neutropenia. This can be congenital (Developed at or before birth) or it can develop later, as in the case of aplastic anemia or some kinds of leukemia. It can also be a side-effect of medication, most prominently chemotherapy. Neutropenia makes an individual highly susceptible to infections. Neutropenia can be the result of colonization by intracellular neutrophilic parasites.

In alpha 1-antitrypsin deficiency, the important neutrophil enzyme elastase is not adequately inhibited by alpha 1-antitrypsin, leading to excessive tissue damage in the presence of inflammation – the most prominent one being pulmonary emphysema. In Familial Mediterranean fever (FMF), a mutation in the pyrin (or marenostrin) gene, which is expressed mainly in neutrophil granulocytes, leads to a constitutively active acute-phase response and causes attacks of fever, arthralgia, peritonitis, and – eventually – amyloidosis.

The Absolute neutrophil count (ANC) is also used in diagnosis and prognosis. The ANC gold standard for determining severity of neutropenia, and thus neutropenic fever. Any ANC < 1500 cells/ mm^3 is considered neutropenia, but <500 cells/ mm^3 is considered severe. There is also new research tying ANC to Myocardial infarction as an aid in early diagnosis.

Neutrophil Antigens

There are five (HNA 1-5) sets of neutrophil antigen recognised. The three HNA-1 antigens (a-c) are located on the low affinity Fc-γ receptor IIIb (FCGR3B :CD16b) The single known HNA-2a antigen is located on CD177. The HNA-3 antigen system has two antigens (3a and 3b) which are located on the seventh exon of the CLT2 gene (SLC44A2). The HNA-4 and HNA-5 antigen systems each have two known antigens (a and b) and are located in the â2 integrin. HNA-4 is located on the áM chain (CD11b) and HNA-5 is located on the áL integrin unit (CD11a).

EOSINOPHIL GRANULOCYTE

Eosinophil granulocytes, usually called eosinophils or eosinophiles (or, less commonly, acidophils), are white blood cells and one of the immune system components responsible for combating multicellular parasites and certain infections in vertebrates. Along with mast cells, they also control mechanisms associated with allergy and asthma. They are granulocytes that develop during hematopoiesis in the bone marrow before migrating into blood.

These cells are eosinophilic or 'acid-loving' as shown by their affinity to coal tar dyes: Normally transparent, it is this affinity that causes them to appear brick-red after staining with eosin, a red dye, using the Romanowsky method. The staining is concentrated in small granules within the cellular cytoplasm, which contain many chemical mediators, such as histamines and proteins such as eosinophil peroxidase, ribonuclease (RNase), deoxyribonucleases, lipase, plasminogen, and major basic protein. These mediators are released by a process called degranulation following activation of the eosinophil, and are toxic to both parasite and host tissues.

In normal individuals, eosinophils make up about 1-6% of white blood cells, and are about 12-17 micrometres in size. They are found in the medulla and the junction between the cortex and medulla of the thymus, and, in the lower gastrointestinal tract, ovary, uterus, spleen, and lymph nodes, but not in the lung, skin, esophagus, or some other internal organs under normal conditions. The presence of eosinophils in these latter organs is associated with disease. Eosinophils persist in the circulation for 8–12 hours, and can survive in tissue for an additional 8–12 days in the absence of stimulation. Pioneering work in the 1980s elucidated that eosinophils were unique granulocytes, having the capacity to survive for extended periods of time after their maturation as demonstrated by ex-vivo culture experiments.

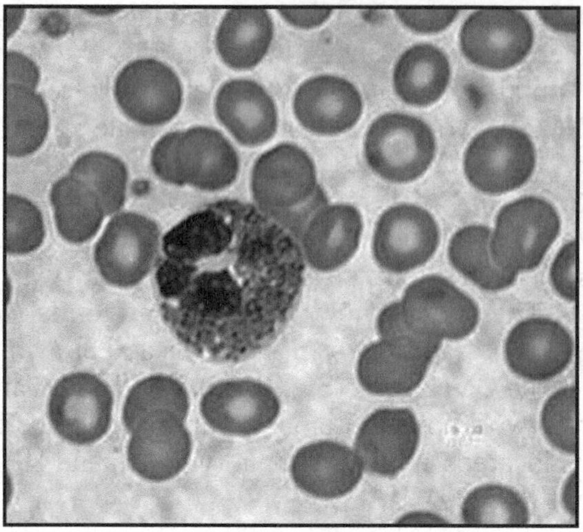

Fig. Eosinophil Under the Microscope (400x) from a Peripheral Blood Smear. Red Blood Cells Surround the Eosinophil, Two Platelets at the Top Left Corner.

Development, Migration and Activation

Eosinophils develop and mature in the bone marrow. They differentiate from myeloid precursor cells in response to the cytokines interleukin 3 (IL-3), interleukin 5 (IL-5), and granulocyte macrophage-colony stimulating factor (GM-CSF). Eosinophils produce and store many secondary granule proteins prior to their exit from the bone marrow. After maturation, eosinophils circulate in blood and migrate to inflammatory sites in tissues, or to sites of helminth infection in response to chemokines like CCL11 (eotaxin-1), CCL24 (eotaxin-2), CCL5 (RANTES), 5-hydroxyicosatetraenoic acid and 5-oxo-eicosatetraenoic acid, and certain leukotrienes like leukotriene B4 (LTB4) and MCP1/4. At these infectious sites, eosinophils are activated by Type 2 cytokines released from a specific subset of helper T cells (T_h2); IL-5, GM-CSF, and IL-3 are important for eosinophil activation as well as maturation. There is evidence to suggest that eosinophil granule protein expression is regulated by the non-coding RNA EGOT (gene)

Function

Function

Following activation, eosinophils effector functions include production of:

- Cationic granule proteins and their release by degranulation.
- Reactive oxygen species such as hypobromite, superoxide, and peroxide (hypobromous acid, which is preferentially produced by eosinophil peroxidase).
- Lipid mediators like the eicosanoids from the leukotriene (e.g., LTC_4, LTD_4, LTE_4) and prostaglandin (e.g., PGE_2) families.
- Enzymes, such as elastase.
- Growth factors such as TGF beta, VEGF, and PDGF.
- Cytokines such as IL-1, IL-2, IL-4, IL-5, IL-6, IL-8, IL-13, and TNF alpha.

In addition, there are also eosinophils that play a role in fighting viral infections, which is evident from the abundance of RNases they contain within their granules, and in fibrin removal during inflammation. Eosinophils along with basophils and mast cells, are important mediators of allergic responses and asthma pathogenesis and are associated with disease severity. They also fight helminth (worm) colonization and may be slightly elevated in the presence of certain parasites. Eosinophils are also involved in many other biological processes, including postpubertal mammary gland development, oestrus cycling, allograft rejection and neoplasia. They have also recently been implicated in antigen presentation to T cells.

Granule Proteins

Following activation by an immune stimulus, eosinophils degranulate to release an array of cytotoxic granule cationic proteins that are capable of inducing tissue damage and dysfunction. These include:

- Major basic protein (MBP)
- Eosinophil cationic protein (ECP)
- Eosinophil peroxidase (EPX)
- Eosinophil-derived neurotoxin (EDN)

Major basic protein, eosinophil peroxidase, and eosinophil cationic protein are toxic to many tissues. Eosinophil cationic protein and eosinophil-derived neurotoxin are ribonucleases with antiviral activity. Major basic protein induces mast cell and basophil degranulation, and is implicated in peripheral nerve remodelling. Eosinophil cationic protein creates toxic pores in the membranes of target cells allowing potential entry of other cytotoxic molecules to the cell, can inhibit proliferation of T cells, suppress antibody production by B cells, induce degranulation by mast cells, and stimulate fibroblast cells to secrete mucus and glycosaminoglycan. Eosinophil peroxidase forms reactive oxygen species and reactive nitrogen intermediates that promote oxidative stress in the target, causing cell death by apoptosis and necrosis.

Eosinophilia

An increase in eosinophils, i.e., the presence of more than 500 eosinophils/ microlitre of blood is called an eosinophilia, and is typically seen in people with a parasitic infestation of the intestines, collagen vascular disease (such as rheumatoid arthritis), malignant diseases such as Hodgkin's disease, extensive skin diseases (such as exfoliative dermatitis), Addison's disease, reflux esophagitis (in which eosinophils will be found in the squamous epithelium of the esophagus), eosinophilic esophagitis, and with the use of certain drugs such as penicillin. But, perhaps the most common cause for eosinophilia is an allergic condition such as asthma. In 1989, contaminated L-tryptophan supplements caused a deadly form of eosinophilia known as eosinophilia-myalgia syndrome, which was reminiscent of the Toxic Oil Syndrome in Spain in 1981.

Treatment

Treatments used to combat autoimmune diseases and conditions caused by eosinophils include:

- Corticosteroids- promote apoptosis. Numbers of eosinophils in blood are rapidly reduced
- Monoclonal antibody therapy- e.g., mepolizumab or reslizumab against IL-5, prevents eosinophilopoiesis
- Antagonists of leukotriene synthesis or receptors
- Gleevec (STI571)- inhibits PDGF-BB in hypereosinophilic leukemia

BASOPHIL GRANULOCYTE

Basophil granulocytes, mostly referred to as basophils, are the least common of the granulocytes, representing about 0.01% to 0.3% of circulating white blood cells. Standard Range is 0.0 - 2.0 % via differential blood count. Basophils contain large cytoplasmic granules which obscure the cell nucleus under the microscope when stained. However, when unstained, the nucleus is visible and it usually has two lobes. The mast cell, another granulocyte, is similar in appearance and function. Both cell types store histamine, a chemical that is secreted by the cells when stimulated. However, they arise from different cell lines, and mast cells usually do not circulate in the blood stream, but instead are located in connective tissue. Like all circulating granulocytes, basophils can be recruited out of the blood into a tissue when needed. The name comes from the fact that these leukocytes are basophilic, i.e., they are susceptible to staining by basic dyes, as shown in the picture.

Function

Basophils appear in many specific kinds of inflammatory reactions, particularly those that cause allergic symptoms. Basophils contain anticoagulant heparin, which prevents blood from clotting too quickly. They also contain the vasodilator histamine, which promotes blood flow to tissues. They can be found in unusually high numbers at sites of ectoparasite infection, e.g., ticks. Like eosinophils, basophils play a role in both parasitic infections and allergies. They are found in tissues where allergic reactions are occurring and probably contribute to the severity of these reactions. Basophils have protein receptors on their cell surface that bind IgE, an immunoglobulin involved in macroparasite defense and allergy.

It is the bound IgE antibody that confers a selective response of these cells to environmental substances, for example, pollen proteins or helminth antigens. Recent studies in mice suggest that basophils may also regulate the behavior of T cells and mediate the magnitude of the secondary immune response.

CD200

Basophil function is inhibited by CD200. Herpesvirus-6, herpesvirus-7, and herpesvirus-8 produce a CD200 homolog which also inhibits basophil function. This suggests that basophils may play a role in the immune response to these viruses.

Immunophenotyping of Basophils

Basophils of mice and humans have consistent immunophenotypes, including FcåRI$^+$, CD123, CD49b(DX-5)$^+$, CD69$^+$, Thy-1.2$^+$, 2B4$^+$, CD11b^{dull}, CD117(c-kit)", CD24", CD19", CD80", CD14", CD23", Ly49c", CD122", CD11c", Gr-1", NK1.1", B220", CD3", γδTCR", αβTCR", α_4 and β_4-integrin negative.

Recently, Heneberg proposed that basophils may be defined as the cellular population positive for CD13, CD44, CD54, CD63, CD69, CD107a, CD123, CD164, CD193/ CCR3, CD203c, TLR-4, and FcåRI. When activated, some additional surface markers are known to be upregulated (CD13, CD107a, CD164), or surface-exposed (CD63, and the ectoenzyme CD203c).

Secretions

Basophils arise and mature in bone marrow. When activated, basophils degranulate to release histamine, proteoglycans (e.g. heparin and chondroitin), and proteolytic enzymes (e.g. elastase and lysophospholipase). They also secrete lipid mediators like leukotrienes (LTD-4), and several cytokines. Histamine and proteoglycans are pre-stored in the cell's granules while the other secreted substances are newly generated. Each of these substances contributes to inflammation. Recent evidence suggests that basophils are an important source of the cytokine, interleukin-4, perhaps more important than T cells. Interleukin-4 is considered one of the critical cytokines in the development of allergies and the production of IgE antibody by the immune system. There are other substances that can activate basophils to secrete which suggests that these cells have other roles in inflammation.

The degranulation of basophils can be investigated in vitro by using flow cytometry and the so-called basophil-activation-test (BAT). Especially, in the diagnosis of allergies including of drug reactions (e.g. induced by contrast medium), the BAT is of great impact. Basopenia (a low basophil count) is difficult to demonstrate as the normal basophil count is so low; it has been reported in association with autoimmune urticaria (a chronic itching condition). Basophilia is also uncommon but may be seen in some forms of leukaemia or lymphoma.

LYMPHOCYTE

A lymphocyte is one of the three subtypes of white blood cell in a vertebrate's immune system. They include natural killer cells (NK cells) (which function in cell-mediated, cytotoxic innate immunity), T cells (for cell-mediated, cytotoxic adaptive immunity), and B cells (for humoral, antibody-driven adaptive immunity). They are the main type of cell found in lymph, which prompted the name lymphocyte.

Types

The three major types of lymphocyte are T cells, B cells and natural killer (NK) cells. Lymphocytes can be identified by their large nucleus.

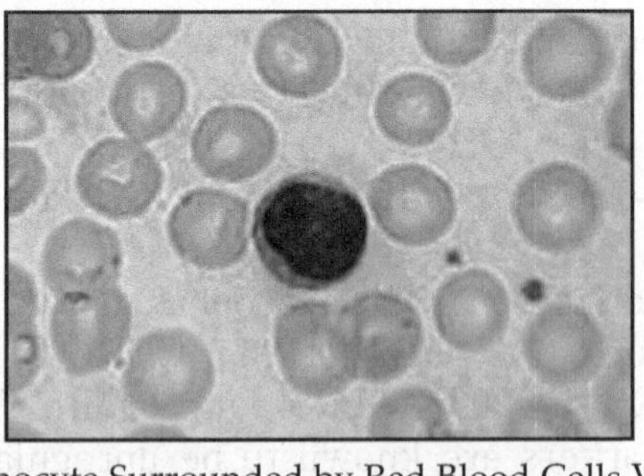

Fig. A Stained Lymphocyte Surrounded by Red Blood Cells Viewed Using a Light Microscope.

T Cells and B Cells

T cells (thymus cells) and B cells (bone marrow- or bursa-derived cells) are the major cellular components of the adaptive immune response. T cells are involved in cell-mediated immunity, whereas B cells are primarily responsible for humoral immunity (relating to antibodies). The function of T cells and B cells is to recognize specific "non-self" antigens, during a process known as antigen presentation. Once they have identified an invader, the cells generate specific responses that are tailored to maximally eliminate specific pathogens or pathogen-infected cells. B cells respond to pathogens by producing large quantities of antibodies which then neutralize foreign objects like bacteria and viruses. In response to pathogens some T cells, called T helper cells, produce cytokines that direct the immune response, while other T cells, called cytotoxic T cells, produce toxic granules that contain powerful enzymes which induce the death of pathogen-infected cells. Following activation, B cells and T cells leave a lasting legacy of the antigens they have encountered, in the form of memory cells. Throughout the lifetime of an animal, these memory cells will "remember" each specific pathogen encountered, and are able to mount a strong and rapid response if the pathogen is detected again.

Natural Killer Cells

NK cells are a part of the innate immune system and play a major role in defending the host from both tumors and virally infected cells. NK cells distinguish infected cells and tumors from normal and uninfected cells by recognizing changes of a surface molecule called MHC (major histocompatibility complex) class I. NK cells are activated in response to a family of cytokines called interferons. Activated NK cells release cytotoxic (cell-killing) granules which then destroy the altered cells. They were named "natural killer cells" because of the initial notion that they do not require prior activation in order to kill cells which are missing MHC class I.

Development

Mammalian stem cells differentiate into several kinds of blood cell within the bone marrow. This process is called haematopoiesis. All lymphocytes originate, during this process, from a common lymphoid progenitor before differentiating into their distinct lymphocyte types. The differentiation of lymphocytes follows various pathways in a hierarchical fashion as well as in a more plastic fashion. The formation of lymphocytes is known as lymphopoiesis. B cells mature into B lymphocytes in the bursa equivalent, which in humans is the GALT, which is thought to be located in the Peyer's patches of the intestine, while T cells migrate to and mature in a distinct organ, called the thymus. Following maturation, the lymphocytes enter the circulation and peripheral lymphoid organs (e.g. the spleen and lymph nodes) where they survey for invading pathogens and/or tumor cells. The lymphocytes involved in adaptive immunity (i.e. B and T cells) differentiate further after exposure to an antigen; they form effector and memory lymphocytes. Effector lymphocytes function to eliminate the antigen, either by releasing antibodies (in the case of B cells), cytotoxic granules (cytotoxic T cells) or by signaling to other cells of the immune system (helper T cells). Memory T cells remain in the peripheral tissues and circulation for an extended time ready to respond to the same antigen upon future exposure.

They live weeks to several years to a whole lifetime, which is very long compared to other leukocytes.

Characteristics

Microscopically, in a Wright's stained peripheral blood smear, a normal lymphocyte has a large, dark-staining nucleus with little to no eosinophilic cytoplasm. In normal situations, the coarse, dense nucleus of a lymphocyte is approximately the size of a red blood cell (about seven micrometres in diameter). Some lymphocytes show a clear perinuclear zone (or halo) around the nucleus or could exhibit a small clear zone to one side of the nucleus. Polyribosomes are a prominent feature in the lymphocytes and can be viewed with an electron microscope. The ribosomes are involved in protein synthesis, allowing the generation of large quantities of cytokines and immunoglobulins by these cells.

It is impossible to distinguish between T cells and B cells in a peripheral blood smear. Normally, flow cytometry testing is used for specific lymphocyte population counts.

This can be used to specifically determine the percentage of lymphocytes that contain a particular combination of specific cell surface proteins, such as immunoglobulins or cluster of differentiation (CD) markers or that produce particular proteins (for example, cytokines using intracellular cytokine staining (ICCS)). In order to study the function of a lymphocyte by virtue of the proteins it generates, other scientific techniques like the ELISPOT or secretion assay techniques can be used.

Typical recognition markers for lymphocytes

Class	Function	Proportion	Phenotypic Marker(S)
NK cells	Lysis of virally infected cells and tumour cells	7% (2-13%)	CD16 CD56 but not CD3
Helper T cells	Release cytokines and growth factors that regulate other immune cells	46% (28-59%)	TCRαβ, CD3 and CD4
Cytotoxic T cells	Lysis of virally infected cells, tumour cells and allografts	19% (13-32%)	TCRαβ, CD3 and CD8
γδ T cells	Immunoregulation and cytotoxicity	5% (2%-8%)	TCRγδ and CD3
B cells	Secretion of antibodies	23% (18-47%)	MHC class II, CD19 and CD21

In the circulatory system, they move from lymph node to lymph node. This contrasts with macrophages, which are rather stationary in the nodes.

Lymphocytes and Disease

A lymphocyte count is usually part of a peripheral complete blood cell count and is expressed as the percentage of lymphocytes to the total number of white blood cells counted. A general increase in the number of lymphocytes is known as lymphocytosis, whereas a decrease is known as lymphocytopenia.

- *High:* An increase in lymphocyte concentration is usually a sign of a viral infection (in some rare case, leukemias are found through an abnormally raised lymphocyte count in an otherwise normal person). A high lymphocyte count with a low neutrophil count might be caused by lymphoma.
- *Low:* A low normal to low absolute lymphocyte concentration is associated with increased rates of infection after surgery or trauma. One basis for low T cell lymphocytes occurs when the human immunodeficiency virus (HIV) infects and destroys T cells (specifically, the $CD4^+$ subgroup of T lymphocytes). Without the key defense that these T cells provide, the body becomes susceptible to opportunistic infections that otherwise would not affect healthy people. The extent of HIV progression is typically determined by measuring the percentage of $CD4^+$ T cells in the patient's blood. The effects of other viruses or lymphocyte disorders can also often be estimated by counting the numbers of lymphocytes present in the blood.
- *Tumor-infiltrating lymphocytes:* In some cancers, e.g., melanoma and colorectal cancer, lymphocytes can migrate into and attack the tumor. This can sometimes lead to regression of the primary tumor.

MONOCYTE

Monocytes are a type of white blood cells (leukocytes). They are the largest of all leukocytes. They are part of the innate immune system of vertebrates including all mammals (humans included), birds, reptiles, and fish. They are amoeboid in shape, having agranulated cytoplasm. Monocytes have unilobar nuclei, which makes them one of the types of mononuclear leukocytes (containing azurophil granules).

The archetypal idea of the nucleus is that it is bean-shaped or kidney-shaped, although the most important distinction is that it is not deeply furcated into lobes, as occurs in polymorphonuclear leukocytes. Monocytes constitute 2% to 10% of all leukocytes in the human body. They play multiple roles in immune function. Such roles include: (1) replenishing resident macrophages under normal states, and (2) in response to inflammation signals, monocytes can move quickly (approx. 8–12 hours) to sites of infection in the tissues and divide/differentiate into macrophages and dendritic cells to elicit an immune response. Half of them are stored in the spleen. Monocytes are usually identified in stained smears by their large kidney shaped or notched nucleus. These change into macrophages after entering into the tissue spaces, and in endothelium can transform into foam cells.

Physiology

Monocytes are produced by the bone marrow from precursors called monoblasts, bipotent cells that differentiated from hematopoietic stem cells. Monocytes circulate in the bloodstream for about one to three days and then typically move into tissues throughout the body. They constitute between three to eight percent of the leukocytes in the blood. Half of them are stored as a reserve in the spleen in clusters in the red pulp's Cords of Billroth. In the tissues, monocytes mature into different types of macrophages at different anatomical locations. Monocytes are the largest corpuscles in the blood.

Monocytes which migrate from the bloodstream to other tissues will then differentiate into tissue resident macrophages or dendritic cells. Macrophages are responsible for protecting tissues from foreign substances, but are also suspected to be important in the formation of important organs like the heart and brain. They are cells that possess a large smooth nucleus, a large area of cytoplasm, and many internal vesicles for processing foreign material.

Monocytes and their macrophage and dendritic-cell progeny serve three main functions in the immune system. These are phagocytosis, antigen presentation, and cytokine production. Phagocytosis is the process of uptake of microbes and particles followed by digestion and destruction of this material. Monocytes can perform phagocytosis using intermediary (opsonising) proteins such as antibodies or complement that coat the pathogen, as well as by binding to the microbe directly via pattern-recognition receptors that recognize pathogens. Monocytes are also capable of killing infected host cells via antibody-dependent cell-mediated cytotoxicity. Vacuolization may be present in a cell that has recently phagocytized foreign matter.

Many factors produced by other cells can regulate the chemotaxis and other functions of monocytes. These factors include most particularly chemokines such as monocyte chemotactic protein-1 and monocyte chemotactic protein-3; certain arachidonic acid metabolites such as Leukotriene B4 and members of the 5-Hydroxyicosatetraenoic acid and 5-oxo-eicosatetraenoic acid family of OXE1 receptor agonists (e.g., 5-HETE and 5-oxo-ETE); and N-Formylmethionine leucyl-phenylalanine and other N-formylated oligopeptides which are made by bacteria and activate the formyl peptide receptor 1.

Microbial fragments that remain after such digestion can serve as antigens. The fragments can be incorporated into MHC molecules and then trafficked to the cell surface of monocytes (and macrophages and dendritic cells). This process is called antigen presentation and it leads to activation of T lymphocytes, which then mount a specific immune response against the antigen.

Other microbial products can directly activate monocytes and this leads to production of pro-inflammatory and, with some delay, of anti-inflammatory cytokines. Typical cytokines produced by monocytes are TNF, IL-1, and IL-12.

Monocyte Subpopulations

There are at least three types of monocytes in human blood:

1. The classical monocyte is characterized by high level expression of the CD14 cell surface receptor (CD14++ CD16- monocyte)
2. The non-classical monocyte shows low level expression of CD14 and additional co-expression of the CD16 receptor (CD14+CD16++ monocyte).
3. The intermediate monocyte with high level expression of CD14 and low level expression of CD16 (CD14++CD16+ monocytes).

Ghattas et al. state that the "intermediate" monocyte population is likely to be a unique subpopulation of monocytes, as opposed to a developmental step, due to their comparatively high expression of surface receptors involved in reparative processes(including vascular endothelial growth factors type 1 and 2, CXCR4, and Tie-2) as well as evidence that the "intermediate" subset is specifically enriched in the bone marrow. After stimulation with microbial products the CD14+CD16++ monocytes produce high amounts of pro-inflammatory cytokines like tumor necrosis factor and interleukin-12.

Said et al. showed that activated monocytes express high levels of PD-1 which might explain the higher expression of PD-1 in CD14+CD16++ monocytes as compared to CD14++CD16- monocytes. Triggering monocytes-expressed PD-1 by its ligand PD-L1 induces IL-10 production which activates CD4 Th2 cells and inhibits CD4 Th1 cell function.

Diagnostic Use

A monocyte count is part of a complete blood count and is expressed either as a ratio of monocytes to the total number of white blood cells counted, or by absolute numbers. Both may be useful in determining or refuting a possible diagnosis.

Monocytosis

Monocytosis is the state of excess monocytes in the peripheral blood. It may be indicative of various disease states. Examples of processes that can increase a monocyte count include:

- Chronic inflammation
- Stress response

- Cushing's syndrome (hyperadrenocorticism)
- Immune-mediated disease
- Pyogranulomatous disease
- Necrosis
- Red blood cell regeneration
- Viral fever
- sarcoidosis

A high count of CD14+CD16++ monocytes is found in severe infection (sepsis) and a very low count of these cells is found after therapy with immuno-suppressive glucocorticoids.

Monocytopenia

Monocytopenia is a form of leukopenia associated with a deficiency of monocytes.

Dendritic Cells

Monocytes can be used to generate dendritic cells in vitro by adding cytokines like Granulocyte Monocyte Colony Stimulating Factor (GM-CSF) and IL-4.

FIXED LEUKOCYTES

Some leukocytes migrate into the tissues of the body to take up a permanent residence at that location rather than remaining in the blood. Often these cells have specific names depending upon which tissue they settle in, such as fixed macrophages in the liver, which become known as Kupffer cells. These cells still serve a role in the immune system.

- Histiocytes
- Dendritic cells (Although these will often migrate to local lymph nodes upon ingesting antigens)
- Mast cells
- Microglia

HISTIOCYTE

A histiocyte is an animal cell that is part of the mononuclear phagocyte system (also known as the reticuloendothelial system or lymphoreticular system). The mononuclear phagocytic system is part of the organism's immune system. The histiocyte is a tissue macrophage or a dendritic cell (histio, diminutive of histo, meaning tissue, and cyte, meaning cell).

Terminology

Histiocytes are derived from the bone marrow by multiplication from a stem cell. The derived cells migrate from the bone marrow to the blood as monocytes. They circulate through the body and enter various organs, where they undergo differentiation into histiocytes, which are part of the mononuclear phagocytic system (MPS).

However, the term histiocyte has been used for multiple purposes in the past, and some cells called "histocytes" do not appear to derive from monocytic-macrophage lines. (The term Histiocyte can also simply refer to a cell from monocyte origin outside the blood system, such as in a tissue (as in rheumatoid arthritis as palisading histiocytes surrounding fibrinoid necrosis of rheumatoid nodules).

Some sources consider Langerhans cell derivatives to be histiocytes. The Langerhans cell histiocytosis embeds this interpretation into its name.

Morphology and Diversity

Histiocytes have common histological and immunophenotypical characteristics (demonstrated by immunostains). Their cytoplasm is eosinophilic and contains variable amounts of lysosomes. They bear membrane receptors for opsonins, such as IgG and the fragment C3b of complement. They express LCAs (leucocyte common antigens) CD45, CD14, CD33, and CD4 (also expressed by T Helper Cells).

Macrophages and Dendritic Cells

These histiocytes are part of the immune system by way of two distinct functions: phagocytosis and antigen presentation. Phagocytosis is the main process of macrophages and antigen presentation the main property of dendritic cells (so called because of their star-like cytoplasmic processes).

Macrophages and dendritic cells are derived from common bone marrow precursor cells that have undergone different differentiation (as histiocytes) under the influence of various environmental (tissue location) and growth factors such as GM-CSF, TNF and IL-4. The various categories of histocytes are distinguishable by their morphology, phenotype, and size.

- Macrophages are highly variable in size and morphology, their cytoplasm contains numerous acid phosphatase laden lysosomes - in relation to their specialised phagocytic function. They express CD68.
- Dendritic cells have an indented (bean-shaped) nucleus and cytoplasm with thin processes (dendritic). Their main activity is antigen presentation; they express Factor XIIIa, CD1c, and Class II Human leukocyte antigens.

Langerhans Cells

A subset of cells differentiates into Langerhans cells; this maturation occurs in the squamous epithelium, lymph nodes, spleen, and bronchiolar epithelium. Langerhans cells are antigen-presenting cells but have undergone further differentiation. Skin Langerhans cells express CD1a, as do cortical thymocytes (cells of the cortex of the thymus gland). They also express S-100, and their cytoplasm contains tennis-racket like ultra-structural inclusions called Birbeck granules.

Disorders

Histiocytoses describe neoplasias wherein the proliferative cell is the histiocyte. The most common histiocyte disorders are Langerhans' cell histiocytosis and haemophagocytic lymphohistiocytosis.

DENDRITIC CELL

Dendritic cells (DCs) are antigen-presenting cells (also known as accessory cells) of the mammalian immune system. Their main function is to process antigen material and present it on the cell surface to the T cells of the immune system. They act as messengers between the innate and the adaptive immune systems.

Dendritic cells are present in those tissues that are in contact with the external environment, such as the skin (where there is a specialized dendritic cell type called the Langerhans cell) and the inner lining of the nose, lungs, stomach and intestines. They can also be found in an immature state in the blood. Once activated, they migrate to the lymph nodes where they interact with T cells and B cells to initiate and shape the adaptive immune response. At certain development stages they grow branched projections, the dendrites that give the cell its name (äÝíäñïí or déndron being Greek for "tree"). While similar in appearance, these are structures distinct from the dendrites of neurons. Immature dendritic cells are also called veiled cells, as they possess large cytoplasmic 'veils' rather than dendrites.

Dendritic cells were first described by Paul Langerhans (hence "Langerhans cells") in the late nineteenth century. The term "dendritic cells" was coined in 1973 by Ralph M. Steinman and Zanvil A. Cohn. For discovering the central role of dendritic cells in the adaptive immune response, Steinman was awarded the Albert Lasker Award for Basic Medical Research in 2007 and the Nobel Prize in Physiology or Medicine in 2011.

Types of Dendritic Cells

As in all dendritic cells, their particular morphology results in a very large contact surface to their surroundings compared to overall cell volume.

In Vivo – Primate

The most common division of dendritic cells is "myeloid" vs. "plasmacytoid": NameDescriptionSecretionToll-like receptorsConventional dendritic cell (previously called Myeloid dendritic cell) (cDC or mDC)Most similar to monocytes. mDC are made up of at least two subsets:(1) the more common mDC-1, which is a major stimulator of T cells(2) the extremely rare mDC-2, which may have a function in fighting wound infectionInterleukin 12 (IL-12)TLR 2, TLR 4Plasmacytoid dendritic cell (pDC)Look like plasma cells, but have certain characteristics similar to myeloid dendritic cells.Can produce high amounts of interferon-alpha and were previously called interferon-producing cells.TLR 7, TLR 9The markers BDCA-2, BDCA-3, and BDCA-4 can be used to discriminate among the types.

Lymphoid and myeloid DCs evolve from lymphoid and myeloid precursors, respectively, and thus are of hematopoietic origin. By contrast, follicular dendritic cells (FDC) are probably of mesenchymal rather than hematopoietic origin and do not express MHC class II, but are so named because they are located in lymphoid follicles and have long "dendritic" processes.

Dendritic Cells in Blood

The blood DCs are typically identified and enumerated in flow cytometry. Three types of DCs have been defined in human blood: the CD1c+ myeloid DCs, the CD141+ myeloid DCs and the CD303+ plasmacytoid DCs. This represents the nomenclature proposed by the nomenclature committee of the International Union of Immunological Societies. Dendritic cells that circulate in blood do not have all the typical features of their counterparts in tissue, i.e. they are less mature and have no dendrites. Still, they can perform complex functions including chemokine-production (in CD1c+ myeloid DCs), cross-presentation (in CD141+ myeloid DCs), and IFNalpha production (in CD303+ plasmacytoid DCs).

In Vitro

In some respects, dendritic cells cultured in vitro do not show the same behaviour or capability as dendritic cells isolated ex vivo. Nonetheless, they are often used for research as they are still much more readily available than genuine DCs.

- Mo-DC or MDDC refers to cells matured from monocytes
- HP-DC refers to cells derived from hematopoietic progenitor cells.

Nonprimate

While humans and non-human primates such as rhesus macaques appear to have DCs divided into these groups, other species (such as the mouse) have different subdivisions of DCs.

Life cycle

Formation of Immature Cells and their Maturation

Dendritic cells are derived from hematopoietic bone marrow progenitor cells. These progenitor cells initially transform into immature dendritic cells.

These cells are characterized by high endocytic activity and low T-cell activation potential. Immature dendritic cells constantly sample the surrounding environment for pathogens such as viruses and bacteria. This is done through pattern recognition receptors (PRRs) such as the toll-like receptors (TLRs). TLRs recognize specific chemical signatures found on subsets of pathogens.

Immature dendritic cells may also phagocytize small quantities of membrane from live own cells, in a process called nibbling. Once they have come into contact with a presentable antigen, they become activated into mature dendritic cells and begin to migrate to the lymph node. Immature dendritic cells phagocytose pathogens and degrade their proteins into small pieces and upon maturation present those fragments at their cell surface using MHC molecules.

Simultaneously, they upregulate cell-surface receptors that act as co-receptors in T-cell activation such as CD80 (B7.1), CD86 (B7.2), and CD40 greatly enhancing their ability to activate T-cells. They also upregulate CCR7, a chemotactic receptor that induces the dendritic cell to travel through the blood stream to the spleen or through the lymphatic system to a lymph node. Here they act as antigen-presenting cells: they activate helper T-cells and killer T-cells as well as B-cells by presenting them with antigens derived from the pathogen, alongside non-antigen specific costimulatory signals. Dendritic cells can also induce T-cell tolerance (unresponsiveness). Certain C-type lectin receptors (CLRs) on the surface of dendritic cells, some functioning as PRRs, help instruct dendritic cells as to when it is appropriate to induce immune tolerance rather than lymphocyte activation.

Every helper T-cell is specific to one particular antigen. Only professional antigen-presenting cells (macrophages, B lymphocytes, and dendritic cells) are able to activate a resting helper T-cell when the matching antigen is presented. However, macrophages and B cells can only activate memory T cells whereas dendritic cells can activate both memory and naive T cells, and are the most potent of all the antigen-presenting cells.

Whereas mature dendritic cells are able to activate antigen-specific naive $CD8^+$ T cells, the formation of $CD8^+$ memory T cells requires the interaction of dendritic cells with $CD4^+$ helper T cells. This help from $CD4^+$ T cells additionally activates the matured dendritic cells and licenses them to efficiently induce $CD8^+$ memory T cells, which are also able to be expanded a second time. For this activation of dendritic cells, concurrent interaction of all three cell types, namely $CD4^+$ T helper cells, $CD8^+$ T cells and dendritic cells, seems to be required.

As mentioned above, mDC probably arise from monocytes, white blood cells which circulate in the body and, depending on the right signal, can turn into either dendritic cells or macrophages. The monocytes in turn are formed from stem cells in the bone marrow. Monocyte-derived dendritic cells can be generated in vitro from peripheral blood mononuclear cells (PBMCs). Plating of PBMCs in a tissue culture flask permits adherence of monocytes. Treatment of these monocytes with interleukin 4 (IL-4) and granulocyte-macrophage colony stimulating factor (GM-CSF) leads to differentiation to immature dendritic cells (iDCs) in about a week. Subsequent treatment with tumor necrosis factor (TNF) further differentiates the iDCs into mature dendritic cells.Monocytes can be induced to differentiate into dendritic cells by a self-peptide Ep1.B derived from apolipoprotein E. These are primarily tolerogenic plasmacytoid dendritic cells.

Life Span of Dendritic Cells

Activated macrophages have a lifespan of only a few days though new evidence suggest that it could be extended to weeks rather than days. The lifespan of activated dendritic cells, while somewhat varying according to type and origin, is of a similar order of magnitude, but immature dendritic cells seem to be able to exist in an inactivated state for much longer.

Research Challenges

The exact genesis and development of the different types and subsets of dendritic cells and their interrelationship is only marginally understood at the moment, as dendritic cells are so rare and difficult to isolate that only in recent years they have become subject of focused research. Distinct surface antigens that characterize dendritic cells have only become known from 2000 on; before that, researchers had to work with a 'cocktail' of several antigens which, used in combination, result in isolation of cells with characteristics unique to DCs.

Dendritic Cells and Cytokines

The dendritic cells are constantly in communication with other cells in the body. This communication can take the form of direct cell–cell contact based on the interaction of cell-surface proteins. An example of this includes the interaction of the membrane proteins of the B7 family of the dendritic cell with CD28 present on the lymphocyte. However, the cell–cell interaction can also take place at a distance via cytokines.

For example, stimulating dendritic cells in vivo with microbial extracts causes the dendritic cells to rapidly begin producing IL-12. IL-12 is a signal that helps send naive CD4 T cells towards a Th1 phenotype. The ultimate consequence is priming and activation of the immune system for attack against the antigens which the dendritic cell presents on its surface. However, there are differences in the cytokines produced depending on the type of dendritic cell. The plasmacytoid DC has the ability to produce huge amounts of type-1 IFN's, which recruit more activated macrophages to allow phagocytosis.

Disease

HIV Infection

HIV, which causes AIDS, can bind to dendritic cells via various receptors expressed on the cell. The best studied example is DC-SIGN (usually on MDC subset 1, but also on other subsets under certain conditions; since not all dendritic cell subsets express DC-SIGN, its exact role in sexual HIV-1 transmission is not clear). When the dendritic cell takes up HIV and then travels to the lymph node, the virus can be transferred to helper CD4+ T-cells, contributing to the developing infection. This infection of dendritic cells by HIV explains one mechanism by which the virus could persist after prolonged HAART. Many other viruses, such as the SARS virus seems to use DC-SIGN to 'hitchhike' to its target cells. However, most work with virus binding to DC-SIGN expressing cells has been conducted using in vitro derived cells such as moDCs. The physiological role of DC-SIGN in vivo is more difficult to ascertain.

Autoimmunity

Altered function of dendritic cells is also known to play a major or even key role in allergy and autoimmune diseases like lupus erythematosus and inflammatory bowel diseases (Crohn's disease and ulcerative colitis).

Dendritic Cells in Animals Other than Humans

The above applies to humans. In other organisms, the function of dendritic cells can differ slightly. For example, in brown rats (but not mice), a subset of dendritic cells exists that displays pronounced killer cell-like activity, apparently through its entire lifespan. However, the principal function of dendritic cells as known to date is always to act as an immune sentinel. They survey the body and collect information relevant to the immune system, they are then able to instruct and direct the adaptive arms to respond to challenges.

In addition, an immediate precursor to myeloid and lymphoid dendritic cells of the spleen has been identified. This precursor, termed pre-DC, lacks MHC class II surface expression, and is distinct from monocytes, which primarily give rise to DCs in non-lymphoid tissues. Dendritic cells have also been found in chickens and turtles.

MAST CELL

A mast cell (also known as a mastocyte or a labrocyte) is derived from the myeloid stem cell. It is a part of the immune system and contains many granules rich in histamine and heparin. Although best known for their role in allergy and anaphylaxis, mast cells play an important protective role as well, being intimately involved in wound healing, including angiogenesis, and defense against pathogens. The mast cell is very similar in both appearance and function to the basophil, another type of white blood cell. They differ in that mast cells are tissue resident, e.g., in mucosal tissues, while basophils are found in the blood.

Mast cells were first described by Paul Ehrlich in his 1878 doctoral thesis on the basis of their unique staining characteristics and large granules. These granules also led him to the incorrect belief that they existed to nourish the surrounding tissue, so he named them Mastzellen (from German Mast, meaning "fattening", as of animals). They are now considered to be part of the immune system.

Mast cells are very similar to basophil granulocytes (a class of white blood cells) in blood. Both are granulated cells that contain histamine and heparin, an anticoagulant. Both cells also release histamine upon binding to immunoglobulin E. These similarities have led many to speculate that mast cells are basophils that have "homed in" on tissues. Furthermore they share a common precursor in bone marrow expressing the CD34 molecule. Basophils leave the bone marrow already mature, whereas the mast cell circulates in an immature form, only maturing once in a tissue site. The site an immature mast cell settles in probably determines its precise characteristics. The first in vitro differentiation and growth of a pure population of mouse mast cells has been carried out using conditioned medium derived from concanavalin A-stimulated splenocytes. Later, it was discovered that T cell-derived interleukin 3 was the component present in the conditioned media that was required for mast cell differentiation and growth.

Mast cells in rodents are classically divided into two subtypes: connective tissue-type mast cells and mucosal mast cells. The activities of the latter are dependent on T-cells.

Mast cells are present in most tissues characteristically surrounding blood vessels and nerves, and are especially prominent near the boundaries between the outside world and the internal milieu, such as the skin, mucosa of the lungs, and digestive tract, as well as the mouth, conjunctiva, and nose.

PHYSIOLOGY

Mast cells play a key role in the inflammatory process. When activated, a mast cell rapidly releases its characteristic granules and various hormonal mediators into the interstitium. Mast cells can be stimulated to degranulate by direct injury (e.g., physical or chemical [such as opioids, alcohols, and certain antibiotics such as polymyxins]), cross-linking of immunoglobulin E (IgE) receptors, or complement proteins.

Mast cells express a high-affinity receptor (FcåRI) for the Fc region of IgE, the least-abundant member of the antibodies. This receptor is of such high affinity that binding of IgE molecules is in essence irreversible. As a result, mast cells are coated with IgE, which is produced by plasma cells (the antibody-producing cells of the immune system). IgE molecules, like all antibodies, are specific to one particular antigen.

In allergic reactions, mast cells remain inactive until an allergen binds to IgE already coated upon the cell. Other membrane activation events can either prime mast cells for subsequent degranulation or act in synergy with FcåRI signal transduction. In general, allergens are proteins or polysaccharides. The allergen binds to the antigen-binding sites, which are situated on the variable regions of the IgE molecules bound to the mast cell surface. It appears that binding of two or more IgE molecules (cross-linking) is required to activate the mast cell. The clustering of the intracellular domains of the cell-bound Fc receptors, which are associated with the cross-linked IgE molecules, causes a complex sequence of reactions inside the mast cell that lead to its activation. Although this reaction is most well-understood in terms of allergy, it appears to have evolved as a defense system against parasites and bacteria. The molecules released into the extracellular environment include:

- Preformed mediators (from the granules):
  - Serine proteases, such as tryptase
  - Histamine (2-5 pg/cell)
  - Serotonin
  - Proteoglycans, mainly heparin (active as anticoagulant)
- Newly formed lipid mediators (eicosanoids):
  - Thromboxane
  - Prostaglandin D2
  - Leukotriene C4
  - Platelet-activating factor
- Cytokines
  - Eosinophil chemotactic factor

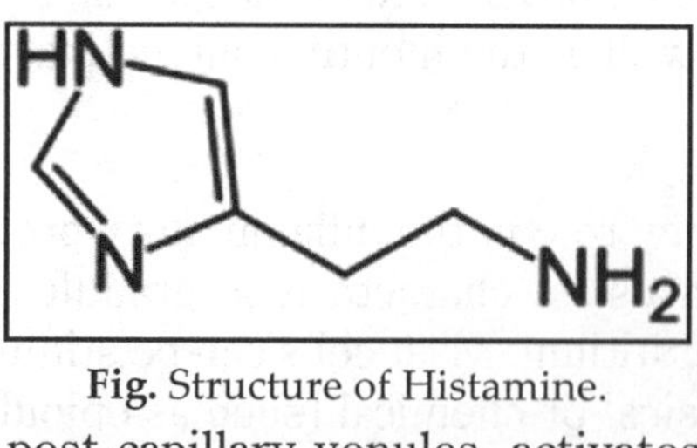

Fig. Structure of Histamine.

Histamine dilates post-capillary venules, activates the endothelium, and increases blood vessel permeability. This leads to local edema (swelling), warmth, redness, and the attraction of other inflammatory cells to the site of release. It also depolarizes nerve endings (leading to itching or pain). Cutaneous signs of histamine release are the "flare and wheal"-reaction. The bump and redness immediately following a mosquito bite are a good example of this reaction, which occurs seconds after challenge of the mast cell by an allergen. The other physiologic activities of mast cells are much less-understood. Several lines of evidence suggest that mast cells may have a fairly fundamental role in innate immunity: They are capable of elaborating a vast array of important cytokines and other inflammatory mediators such as TNFa; they express multiple "pattern recognition receptors" thought to be involved in recognizing broad classes of pathogens; and mice without mast cells seem to be much more susceptible to a variety of infections.

Mast cell granules carry a variety of bioactive chemicals. These granules have been found to be transferred to adjacent cells of the immune system and neurons in a process of transgranulation via mast cell pseudopodia.

Biochemistry

Structure of FcεR1

FcåR1 is a high affinity IgE-receptor that is expressed on the surface of the mast cell. FcåR1 is a tetramer made of one alpha (α) chain, one beta (β) chain, and two identical, disulfide-linked gamma (γ) chains. The binding site for the IgE is formed by the extracellular portion of the á chain that contains two domains that are similar to Ig. One transmembrane domain contains an aspartic acid residue, and one contains a short cytoplasmic tail. The β chain contains, a single immunoreceptor tyrosine-based activation motif ITAM, in the cytoplasmic region.

Each γ chain has one ITAM on the cytoplasmic region. The signaling cascade from the receptor is initiated when the ITAMs of the β and γ chains are phosphorylated by tyrosine. This signal is required for the activation of mast cells. Type 2 helper T cells,(Th2) and many other cell types lack the β chain, so signaling is mediated only by the γ chain. This is due to the α chain containing endoplasmic reticulum retention signals that causes the α-chains to remain degraded in the ER. The assembly of the á chain with the co-transfected β and γ chains mask the ER retention and allows the α β γ complex to be exported to the golgi apparatus to the plasma membrane in rats. In humans, only the γ complex is needed to counterbalance the á chain ER retention.

Allergen Process

Allergen-mediated FcεR1 cross-linking signals are very similar to the signaling event resulting in antigen binding to lymphocytes. The Lyn tyrosine kinase is associated with the cytoplasmic end of the FcεR1 β chain. The antigen cross-links the FcεR1 molecules, and Lyn tyrosine kinase phosphorylates the ITAMs in the FcεR1 β and γ chain in the cytoplasm. Upon phosphorylation, the Syk tyrosine kinase gets recruited to the ITAMs located on the γ chains. This causes activation of the Syk tyrosine kinase, causing it to phosphorylate. Syk functions as a signal amplifying kinase activity due to the fact that it targets multiple proteins and causes their activation. This antigen stimulated phosphorylation causes the activation of other proteins in the FcåR1-mediated signaling cascade.

Degranulation and Fusion

An important adaptor protein activated by the Syk phosphorylation step is the linker for activation of T cells (LAT). LAT is important because it can be modified by phosphorylation to create novel binding sites.

Phospholipase C (PLCã), becomes phosphorylated once bound to LAT, and is then used to catalyze Phosphatidylinositol bisphosphate breakdown to yield inositol trisphosphate (IP3) and diacyglycerol (DAG). IP3 elevates calcium levels, and DAG activates protein kinase C (PKC).

This is not the only way that PKC is made. The tyrosine kinase, Fyn, phosphorylates a Grb-2-associated binder-like protein 2 (Gab2) which binds to phosphoinositide 3-kinase which activates PKC.PKC leads to the activation of myosin light-chain phosphorylation granule movements which disassembles the actin-myosin complexes to allow granules to come into contact with the plasma membrane.

The mast cell granule can now fuse with the plasma membrane. Soluble N-ethylmaleimide sensitive fusion Attachment Protein Rceptor"SNARE complex mediates this process. Different SNARE proteins interact to form different complexes that catalyze fusion. Rab3 guanosine triphosphatases and Rab-associated kinases and phosphatases regulate the cell granule membrane fusion in resting mast cells.

Enzymes

Enzyme	Function
Lyn tyrosine kinase	Phosphorylates the ITAMs in the FcεR1 β and γ chain in the cytoplasm. It causes Syk tyrosine kinase to get recruited to the ITAMS located on the γ chains. This causes activation of the Syk tyrosine kinase, causing it to phosphorylate
Syk tyrosine kinase	Targets multiple proteins and causes their activation
Phospholipase C	Catalyzes Phosphatidylinositol 4,5-bisphosphate
Inositol trisphosphate	Elevates calcium levels
Diacylglycerol	Activates protein kinase C
Fyn	Phosphorylates Grb-2-associated-binder-like-protein
Grb-2-associated-binder-like-protein	Binds to phosphoinositide 3-kinase
Phosphoinositide 3-kinase	Activates protein kinase C
Protein kinase C	Activates myosin light-chain phosphorylation granule movements that disassemble the actin-myosin complexes
Rab-associated kinases and phosphatases	Regulate cell granule membrane fusion in resting mast cells

Role in Disease

Allergic Disease

Many forms of cutaneous and mucosal allergy are mediated in large part by mast cells; they play a central role in asthma, eczema, itch (from various causes), and allergic rhinitis and allergic conjunctivitis. Antihistamine drugs act by blocking histamine action on nerve endings. Cromoglicate-based drugs (sodium cromoglicate, nedocromil) block a calcium channel essential for mast cell degranulation, stabilizing the cell and preventing release of histamine and related mediators. Leukotriene antagonists (such as montelukast and zafirlukast) block the action of leukotriene mediators and are being used increasingly in allergic diseases.

Calcium triggers the secretion of histamine from mast cells after previous exposure to sodium fluoride. The secretory process can be divided into a fluoride-activation step and a calcium-induced secretory step. It was observed that the fluoride-activation step is accompanied by an elevation of cAMP levels within the cells. The attained high levels of cAMP persist during histamine release. It was further found that catecholamines do not markedly alter the fluoride-induced histamine release. It was also confirmed that the second, but not the first, step in sodium fluoride-induced histamine secretion is inhibited by theophylline. Vasodilation and increased permeability of capillaries are a result of both H1 and H2 receptor types. Stimulation of histamine activates a histamine (H2)-sensitive adenylate cyclase of oxyntic cells, and there is a rapid increase in cellular [cAMP] that is involved in activation of $H+$ transport and other associated changes of oxyntic cells.

Anaphylaxis

In anaphylaxis (a severe systemic reaction to allergens, such as nuts, bee stings, or drugs), body-wide degranulation of mast cells leads to vasodilation and, if severe, symptoms of life-threatening shock. Histamine is a vasodilatory substance released during anaphylaxis.

Autoimmunity

Mast cells may be implicated in the pathology associated with autoimmune, inflammatory disorders of the joints. They have been shown to be involved in the recruitment of inflammatory cells to the joints (e.g., rheumatoid arthritis) and skin (e.g., bullous pemphigoid), and this activity is dependent on antibodies and complement components.

Reproductive Disorders

Mast cells are present within the endometrium, with increased activation and release of mediators in endometriosis. In males, mast cells are present in the testes and are increased in oligo- and azoospermia, with mast cell mediators directly suppressing sperm motility in a potentially reversible manner.

Mast Cell Disorders

Mastocytosis is a rare condition featuring proliferation of mast cells. It exists in a cutaneous and systemic form, with the former being limited to the skin and the latter involving multiple organs. Mast cell tumors are often seen in dogs and cats.

Autism

Research into an immunological contribution to autism suggests that autism spectrum disorder (ASD) children may present with "allergic-like" problems in the absence of elevated serum IgE and chronic urticaria, suggesting non-allergic mast cell activation in response to environmental and stress triggers. This mast cell activation could contribute to brain inflammation and neurodevelopmental problems.

MICROGLIA

Microglia are a type of glial cell that are the resident macrophages of the brain and spinal cord, and thus act as the first and main form of active immune defense in the central nervous system (CNS).

Microglia constitute 10-15% of all cells found within the brain. Microglia (and astrocytes) are distributed in large non-overlapping regions throughout the brain and spinal cord. Microglia are constantly scavenging the CNS for plaques, damaged neurons and infectious agents. The brain and spinal cord are considered "immune privileged" organs in that they are separated from the rest of the body by a series of endothelial cells known as the blood–brain barrier, which prevents most infections from reaching the vulnerable nervous tissue. In the case where infectious agents are directly introduced to the brain or cross the blood–brain barrier, microglial cells must react quickly to decrease inflammation and destroy the infectious agents before they damage the sensitive neural tissue. Due to the unavailability of antibodies from the rest of the body (few antibodies are small enough to cross the blood–brain barrier), microglia must be able to recognize foreign bodies, swallow them, and act as antigen-presenting cells activating T-cells. Since this process must be done quickly to prevent potentially fatal damage, microglia are extremely sensitive to even small pathological changes in the CNS. They achieve this sensitivity in part by having unique potassium channels that respond to even small changes in extracellular potassium.

Microglial cells differentiate in the bone marrow from hematopoietic stem cells, the progenitors of all blood cells. During hematopoiesis, some of these stem cells differentiate into monocytes and travel from the bone marrow to the brain, where they settle and further differentiate into microglia. However, recent studies indicate microglia originate in the yolk sac during a remarkably restricted period and populate the brain mesenchyme.

Monocytes can also differentiate into myeloid dendritic cells and macrophages in the peripheral systems. Like macrophages in the rest of the body, microglia use phagocytic and cytotoxic mechanisms to destroy foreign materials. Microglia and macrophages both contribute to the immune response by acting as antigen presenting cells, as well as promoting inflammation and homeostatic mechanisms within the body by secreting cytokines and other signaling molecules.

In their downregulated form, microglia lack the MHC class I/MHC class II proteins, IFN-γ cytokines, CD45 antigens, and many other surface receptors required to act in the antigen-presenting, phagocytic, and cytotoxic roles that hallmark normal macrophages. Microglia also differ from macrophages in that they are much more tightly regulated spatially and temporally in order to maintain a precise immune response.

Another difference between microglia and other cells that differentiate from myeloid progenitor cells is the turnover rate. Macrophages and dendritic cells are constantly being used up and replaced by myeloid progenitor cells which differentiate into the needed type. Due to the blood–brain barrier, it would be fairly difficult for the body to constantly replace microglia. Therefore, instead of constantly being replaced with myeloid progenitor cells, the microglia maintain their status quo while in their quiescent state, and then, when they are activated, they rapidly proliferate in order to keep their numbers up. Bone chimera studies have shown, however, that in cases of extreme infection the blood–brain barrier will weaken, and microglia will be replaced with haematogenous, marrow-derived cells, namely myeloid progenitor cells and macrophages. Once the infection has decreased the disconnect between peripheral and central systems is reestablished and only microglia are present for the recovery and regrowth period.

The ability to view and characterize different neural cells including microglia began in 1880 when Nissl staining was developed by Franz Nissl. Franz Nissl and F. Robertson first described microglial cells during their histology experiments. The cell staining techniques in the 1880s showed that microglia are related to macrophages. The activation of microglia and formation of ramified microglial clusters was first noted by Victor Babe° while studying a rabies case in 1897. Babe° noted the cells were found in a variety of viral brain infections but did not know what the clusters of microglia he saw were. Pío del Río Hortega, a student of Santiago Ramón y Cajal, first called the cells "microglia" around 1920. He went on to characterize microglial response to brain lesions in 1927 and note the "fountains of microglia" present in the corpus callosum and other perinatal white matter areas in 1932. After many years of research Rio-Hortega became generally considered as the "Father of Microglia." For a long period of time little improvement was made in our knowledge of microglia. Then, in 1988, Hickey and Kimura showed that perivascular microglial cells are bone-marrow derived, and express high levels of MHC class II proteins used for antigen presentation. This confirmed Pio Del Rio-Hortega's postulate that microglial cells functioned similarly to macrophages by performing phagocytosis and antigen presentation.

Types

Microglial cells are extremely plastic, and undergo a variety of structural changes based on their location and current role. This level of plasticity is required to fulfill the vast variety of immunological functions that microglia perform, as well as maintaining homeostasis within the brain. If microglia were not capable of this they would need to be replaced on a regular basis like macrophages, and would not be available to the CNS immune defense on extremely short notice without causing immunological imbalance under normal conditions.

Amoeboid

This form of microglial cell is found mainly within the perinatal white matter areas in the corpus callosum known as the "Fountains of Microglia." This shape allows the microglial free movement throughout the neural tissue, which allows it to fulfill its role as a scavenger cell. Amoeboid microglia are able to phagocytose debris, but do not fulfill the same antigen-presenting and inflammatory roles as activated microglia. Amoeboid microglia are especially prevalent during the development and rewiring of the brain, when there are large amounts of extracellular debris and apoptotic cells to remove.

Ramified

This form of microglial cell is commonly found at specific locations throughout the entire brain and spinal cord in the absence of foreign material or dying cells. This "resting" form of microglia is composed of long branching processes and a small cellular body. Unlike the ameboid forms of microglia, the cell body of the ramified form remains fairly motionless, while its branches are constantly moving and surveying the surrounding area. The branches are very sensitive to small changes in physiological condition and require very specific culture conditions to observe in vitro. Unlike activated or ameboid microglia, ramified microglia are unable to phagocytose cells and display little or no immunomolecules. This includes the MHC class I/II proteins normally used by macrophages and dendritic cells to present antigens to t-cells, and as a result ramified microglia function extremely poorly as antigen presenters. The purpose of this state is to maintain a constant level of available microglia to detect and fight infection, while maintaining an immunologically silent environment.

Activated

Activated microglia can be stained via the marker Iba1, which is upregulated during activation.

Non-Phagocytic

This state is actually part of a graded response as microglia move from their ramified form to their fully active phagocytic form. Microglia can be activated by a variety of factors including: glutamate receptor agonists, pro-inflammatory cytokines, cell necrosis factors, lipopolysaccharide, and changes in extracellular potassium (indicative of ruptured cells). Once activated the cells undergo several key morphological changes including the thickening and retraction of branches, uptake of MHC class I/II proteins, expression of immunomolecules, secretion of cytotoxic factors, secretion of recruitment molecules, and secretion of pro-inflammatory signaling molecules (resulting in a pro-inflammation signal cascade). Activated non-phagocytic microglia generally appear as "bushy," "rods," or small ameboids depending on how far along the ramified to full phagocytic transformation continuum they are. In addition, the microglia also undergo rapid proliferation in order to increase their numbers. From a strictly morphological perspective, the variation in microglial form along the continuum is associated with changing morphological complexity and can be quantitated using the methods of fractal analysis, which have proven sensitive to even subtle, visually undetectable changes associated with different morphologies in different pathological states.

Phagocytic

Activated phagocytic microglia are the maximally immune responsive form of microglia. These cells generally take on a large, ameboid shape, although some variance has been observed. In addition to having the antigen presenting, cytotoxic and inflammatory mediating signaling of activated non-phagocytic microglia, they are also able to phagocytose foreign materials and display the resulting immunomolecules for T-cell activation. Phagocytic microglia travel to the site of the injury, engulf the offending material, and secrete pro-inflammatory factors to promote more cells to proliferate and do the same. Activated phagocytic microglia also interact with astrocytes and neural cells to fight off the infection as quickly as possible with minimal damage to the healthy brain cells.

Gitter Cells

Gitter cells are the eventual result of microglial cell's phagocytosis of infectious material or cellular debris. Eventually, after engulfing a certain amount of material, the phagocytic microglia becomes unable to phagocytose any further materials. The resulting cellular mass is known as a granular corpuscle, named for its 'grainy' appearance. By looking at tissues stained to reveal gitter cells, pathologists can see post-infection areas that have healed.

Perivascular

Unlike the other types of microglia mentioned above, "perivascular" microglia refers to the location of the cell rather than its form/function. Perivascular microglia are mainly found encased within the walls of the basal lamina. They perform normal microglial functions, but unlike normal microglia they are replaced by bone marrow derived precursor cells on a regular basis and express MHC class II antigens regardless of the outside environment. Perivascular microglia also react strongly to macrophage differentiation antigens. These microglia have been shown to be essential to repair of vascular walls, as shown by Ritter's experiments and observations on ischemic retinopathy. Perivascular microglia promote endothelial cell proliferation allowing new vessels to be formed and damaged vessels to be repaired. During repair and development, myeloid recruitment and differentiation into microglial cells is highly accelerated to accomplish these tasks.

Juxtavascular

Like perivascular microglia, juxtavascular microglia can be distinguished mainly by their location. Juxtavascular microglia are found making direct contact with the basal lamina wall of blood vessels but are not found within the walls. Like perivascular cells, they express MHC class II proteins even at low levels of inflammatory cytokine activity. Unlike perivascular cells, but similar to resident microglia, juxtavascular microglia do not exhibit rapid turnover or replacement with myeloid precursor cells on a regular basis.

Functions

Microglial cells fulfill a variety of different tasks within the CNS mainly related to both immune response and maintaining homeostasis. The following are some of the major known functions carried out by these cells.

Scavenging

In addition to being very sensitive to small changes in their environment, each microglial cell also physically surveys its domain on a regular basis. This action is carried out in the ameboid and resting states. While moving through its set region, if the microglial cell finds any foreign material, damaged cells, apoptotic cells, neurofibrillary tangles, DNA fragments, or plaques it will activate and phagocytose the material or cell. In this manner microglial cells also act as "housekeepers" cleaning up random cellular debris. During developmental wiring of the brain, microglial cells play a large role removing unwanted excess cellular matter. Post development, the majority of dead or apoptotic cells are found in the cerebral cortex and the subcortical white matter. This may explain why the majority of ameboid microglial cells are found within the "fountains of microglia" in the cerebral cortex.

Phagocytosis

The main role of microglia, phagocytosis, involves the engulfing of various materials. Engulfed materials generally consist of cellular debris, lipids, and apoptotic cells in the non-inflamed state, and invading virus, bacteria, or other foreign materials in the inflamed state. Once the microglial cell is "full" it stops phagocytic activity and changes into a relatively non-reactive gitter cell.

Cytotoxicity

In addition to being able to destroy infectious organisms through cell to cell contact via phagocytosis, microglia can also release a variety of cytotoxic substances. Microglia in culture secrete large amounts of H_2O_2 and NO in a process known as 'respiratory burst'. Both of these chemicals can directly damage cells and lead to neuronal cell death. Proteases secreted by microglia catabolise specific proteins causing direct cellular damage, while cytokines like IL-1 promote demyelination of neuronal axons.

Finally, microglia can injure neurons through NMDA receptor-mediated processes by secreting glutamate,aspartate and quinolinic acid. Cytotoxic secretion is aimed at destroying infected neurons, virus, and bacteria, but can also cause large amounts of collateral neural damage. As a result, chronic inflammatory response can result in large scale neural damage as the microglia ravage the brain in an attempt to destroy the invading infection.

Antigen Presentation

As mentioned above, resident non-activated microglia act as poor antigen presenting cells due to their lack of MHC class I/II proteins. Upon activation they rapidly uptake MHC class I/II proteins and quickly become efficient antigen presenters. In some cases, microglia can also be activated by IFN-γ to present antigens, but do not function as effectively as if they had undergone uptake of MHC class I/II proteins.

During inflammation, T-cells cross the blood–brain barrier thanks to specialized surface markers and then directly bind to microglia in order to receive antigens. Once they have been presented with antigens, T-cells go on to fulfill a variety of roles including pro-inflammatory recruitment, formation of immunomemories, secretion of cytotoxic materials, and direct attacks on the plasma membranes of foreign cells.

Synaptic Stripping

In a phenomenon first noticed in spinal lesions by Blinzinger and Kreutzberg in 1968, post-inflammation microglia remove the branches from nerves near damaged tissue. This helps promote regrowth and remapping of damaged neural circuitry.

Promotion of Repair

Post-inflammation, microglia undergo several steps to promote regrowth of neural tissue. These include synaptic stripping, secretion of anti-inflammatory cytokines, recruitment of neurons and astrocytes to the damaged area, and formation of gitter cells. Without microglial cells regrowth and remapping would be considerably slower in the resident areas of the CNS and almost impossible in many of the vascular systems surrounding the brain and eyes.

Extracellular Signaling

A large part of microglial cell's role in the brain is maintaining homeostasis in non-infected regions and promoting inflammation in infected or damaged tissue. Microglia accomplish this through an extremely complicated series of extracellular signaling molecules which allow them to communicate with other microglia, astrocytes, nerves, T-cells, and myeloid progenitor cells. As mentioned above the cytokine IFN-γ can be used to activate microglial cells. In addition, after becoming activated with IFN-γ, microglia also release more IFN-γ into the extracellular space. This activates more microglia and starts a cytokine induced activation cascade rapidly activating all nearby microglia. Microglia-produced TNF-γ causes neural tissue to undergo apoptosis and increases inflammation. IL-8 promotes B-cell growth and differentiation, allowing it to assist microglia in fighting infection. Another cytokine, IL-1, inhibits the cytokines IL-10 and TGF-β, which downregulate antigen presentation and pro-inflammatory signaling. Additional dendritic cells and T-cells are recruited to the site of injury through the microglial production of the chemotactic molecules like MDC, IL-8, and MIP-3β. Finally, PGE_2 and other prostanoids prevent chronic inflammation by inhibiting microglial pro-inflammatory response and downregulating Th1 (T-helper cell) response.

Role in Chronic Neuroinflammation

The word neuroinflammation has come to stand for chronic, central nervous system (CNS) specific, inflammation-like glial responses that may produce neurodegenerative symptoms such as plaque formation, dystrophic neurite growth, and excessive tau phosphorylation. It is important to distinguish between acute and chronic neuroinflammation. Acute neuroinflammation is generally caused by some neuronal injury after which microglia migrate to the injured site engulfing dead cells and debris. The term neuroinflammation generally refers to more chronic, sustained injury when the responses of microglial cells contribute to and expand the neurodestructive effects, worsening the disease process.

When microglia are activated they take on an amoeboid shape and they alter their gene expression. Altered gene expression leads to the production of numerous potentially neurotoxic mediators. These mediators are important in the normal functions of microglia and their production is usually decreased once their task is complete. In chronic neuroinflammation, microglia remain activated for an extended period during which the production of mediators is sustained longer than usual. This increase in mediators contributes to neuronal death.

Neuroinflammation is distinct from inflammation in other organs, but does include some similar mechanisms such as the localized production of chemoattractant molecules to the site of inflammation. The following list contains a few of the numerous substances that are secreted when microglia are activated:

Cytokines

Microglia activate the proinflammatory cytokines IL-1α, IL-1β and TNF-α in the CNS. Cytokines play a potential role in neurodegeneration when microglia remain in a sustained activated state. Direct injection of the cytokines IL-1α, IL-1β and TNF-α into the CNS result in local inflammatory responses and neuronal degradation. This is in contrast with the potential neurotrophic (inducing growth of neurons) actions of these cytokines during acute neuroinflammation.

Chemokines

Chemokines are cytokines that stimulate directional migration of inflammatory cells in vitro and in vivo. Chemokines are divided into four main subfamilies: C, CC, CXC, and CX_3C. Microglial cells are sources of some chemokines and express the monocyte chemoattractant protein-1 (MCP-1) chemokine in particular. Other inflammatory cytokines like IL-1β and TNF-α, as well as bacterial-derived lipopolysaccharide (LPS) may stimulate microglia to produce MCP-1, MIP-1α, and MIP-1β. Microglia can express CCR3, CCR5, CXCR4, and CX3CR1 in vitro. Chemokines are proinflammatory and therefore contribute to the neuroinflammation process.

Proteases

When microglia are activated they induce the synthesis and secretion of proteolytic enzymes that are potentially involved in many functions. There are a number of proteases that possess the potential to degrade both the extracellular matrix and neuronal cells that are in the neighborhood of the microglia releasing these compounds. These proteases include; cathepsins B,

L, and S, the matrix metalloproteinases MMP-1, MMP-2, MMP-3, and MMP-9, and the metalloprotease-disintegrin ADAM8 (plasminogen) which forms outside microglia and degrades the extracellular matrix. Both Cathepsin B, MMP-1 and MMP-3 have been found to be increased in Alzheimer's disease (AD) and cathepsin B is increased in multiple sclerosis (MS). Elastase, another protease, could have large negative effects on the extracellular matrix.

Amyloid Precursor Protein

Microglia synthesize amyloid precursor protein (APP) in response to excitotoxic injury. Plaques result from abnormal proteolytic cleavage of membrane bound APP. Amyloid plaques can stimulate microglia to produce neurotoxic compounds such as cytokines, excitotoxin, nitric oxide and lipophylic amines, which all cause neural damage. Plaques in Alzheimer's disease contain activated microglia. A study has shown that direct injection of amyloid into brain tissue activates microglia, which reduces the number of neurons. Microglia have also been suggested as a possible source of secreted â amyloid.

Aging

Microglia undergo a burst of mitotic activity during injury; this proliferation is followed by apoptosis to reduce the cell numbers back to baseline. Activation of microglia places a load on the anabolic and catabolic machinery of the cells causing activated microglia to die sooner than non-activated cells. To compensate for microglial loss over time, microglia undergo mitosis and bone marrow derived progenitor cells migrate into the brain via the meninges and vasculature.

Accumulation of minor neuronal damage that occurs during normal aging can transform microglia into enlarged and activated cells. These chronic, age-associated increases in microglial activation and IL-1 expression may contribute to increased risk of Alzheimer's disease with advancing age through favoring neuritic plaque formation in susceptible patients. DNA damage might contribute to age-associated microglial activation. Another factor might be the accumulation of advanced glycation endproducts, which accumulate with aging. These proteins are strongly resistant to proteolytic processes and promote protein cross-linking.

Research has discovered dystrophic (defective development) human microglia. "These cells are characterized by abnormalities in their cytoplasmic structure, such as deramified, atrophic, fragmented or unusually tortuous processes, frequently bearing spheroidal or bulbous swellings." The incidence of dystrophic microglia increases with aging. Microglial degeneration and death have been reported in research on Prion disease, Schizophrenia and Alzheimer's disease, indicating that microglial deterioration might be involved in neurodegenerative diseases. A complication of this theory is the fact that it is difficult to distinguish between "activated" and "dystrophic" microglia in the human brain.

Role of Microglia in Neurodegeneration

Neurodegenerative disorders are characterized by progressive cell loss in specific neuronal populations. "Many of the normal trophic functions of glia may be lost or overwhelmed when the cells become chronically activated in progressive neurodegenerative disorders, for there is abundant evidence that in such disorders, activated glia play destructive roles by direct and indirect inflammatory attack." The following are prominent examples of microglial cells' role in neurodegenerative disorders.

Alzheimer's Disease

Alzheimer's disease (AD) is a progressive, neurodegenerative disease where the brain develops abnormal clumps (amyloid plaques) and tangled fiber bundles (neurofibrillary tangles).

There are many activated microglia over-expressing IL-1 in the brains of Alzheimer patients that are distributed with both Aβ plaques and neurofibrillary tangles. This over expression of IL-1 leads to excessive tau phosphorylation that is related to tangle development in Alzheimer's disease.

Many activated microglia are found to be associated with amyloid deposits in the brains of Alzheimer's patients. Microglia interact with β-amyloid plaques through cell surface receptors that are linked to tyrosine kinase based signaling cascades that induce inflammation. When microglia interact with the deposited fibrillar forms of β-amyloid it leads to the conversion of the microglia into an activated cell and results in the synthesis and secretion of cytokines and other proteins that are neurotoxic.

One preliminary model as to how this would occur involves a positive feedback loop. When activated, microglia will secrete proteases, cytokines, and reactive oxygen species. The cytokines may induce neighboring cells to synthesize amyloid precursor protein. The proteases then possibly could cause the cleaving required to turn precursor molecules into the beta amyloid that characterizes the disease. Then, the oxygen species encourage the aggregation of beta amyloid in order to form plaques. The growing size of these plaques then in turn triggers the action of even more microglia, which then secrete more cytokines, proteases, and oxygen species, thus amplifying the neurodegeneration.

Treatment

Non-steroidal anti-inflammatory drugs (NSAIDs) have proven to be effective in reducing the risk of AD. "Sustained treatment with NSAIDs lowers the risk of AD by 55%, delays disease onset, attenuates symptomatic severity and slows the loss of cognitive abilities. The main cellular target for NSAIDs is thought to be microglia. This is supported by the fact that in patients taking NSAIDs the number of activated microglia is decreased by 65%."

Parkinson's Disease

Parkinson's disease is a movement disorder in which the dopamine-producing neurons in the brain do not function as they should, the neurons of the Substantia Nigra become dysfunctional and eventually die, leaving a lack of dopamine input into the striatum. This causes the symptoms of Parkinson's disease.

Cardiovascular Diseases

Recently microglial activation has been reported in rats with myocardial infarction (Rana et al.,2010). This activation was specific to brain nuclei involved in cardiovascular regulation suggesting possible role of microglial activation in pathogenesis of heart failure.

Role in Viral Infections

Human Immunodeficiency Virus

The infection of mononuclear phagocytes with HIV-1 is an important element in the development of HIV-associated dementia complex (HAD). The only brain cell type that is "productively" infected with the virus are microglial cells. It has also become clear that neurotoxic mediators released from brain microglia play an important role in the pathogenesis of HIV-1.

"HIV-1 can enter the microglial cell via CD4 receptors and chemokine co-receptors such as CCR3, CCR5, and CXCR4, with CCR5 being the most important of these. Interestingly, humans with double allelic loss of CCR5 are virtually immune to HIV acquired via the sexual route (though can be infected by IV transmission of CXCR4 tropic viruses). IL-4 and IL-10 enhance the entry and replication of HIV-1 in microglia through the up-regulation of CD4 and CCR5 expression, respectively. The chemokines CCL5/RANTES, CCL3/MIP-1α, CCL4/MIP-1β, all of which bind to CCR5, are inhibitory to HIV-1 replication in microglial cells, apparently by their ability to block viral entry."

Infected microglia contain viral particles intracellularly. There is a correlation between the severity of dementia and microglial production of neurotoxins.

One discrepancy in HAD is the limited number of HIV-1 infected microglia in comparison to the many CNS abnormalities that occur. This suggests that chemical factors that are released from microglial cells are contributing to neuronal loss. "It has become more and more apparent that HIV-1 infected microglial cells actively secrete both endogenous neurotoxins such as TNF-α,

IL-1β, CXCL8/IL-8, glutamate, quinolinic acid, platelet activating factor, eicosanoids, and no as well as the neurotoxic viral proteins Tat, gp120, and gp41."

Microglia are the main target of HIV-1 in the brain. When activated by HIV-1 or viral proteins, they secrete or induce other cells to secrete neurotoxic factors; this process is accompanied by neuronal dysfunction (HAD).

Herpes Simplex Virus

Herpes simplex virus (HSV) can cause herpes encephalitis in babies and immunocompetent adults. Studies have shown that long-term neuroimmune activation persists after the herpes infection in patients. Microglia produce cytokines that are toxic to neurons; this may be a mechanism underlying HSV-related CNS damage. It has been found that "active microglial cells in HSV encephalitis patients do persist for more than 12 months after antiviral treatment."

Role in Bacterial Infections

Lipopolysaccharide (LPS) is the major component of the outer membrane of a gram-negative bacterial cell wall. LPS has been shown to activate microglia in vitro and stimulates microglia to produce cytokines, chemokines, and prostaglandins. "Although LPS has been used as a classic activating agent, a recent study of rat microglia demonstrated that prolonged LPS exposure induces a distinctly different activated state from that in microglia acutely exposed to LPS."

Streptococcus Pneumoniae

Streptococcus pneumoniae is the most common cause of bacterial meningitis. It is primarily localized to the subarachnoid space while cytokines and chemokines are produced inside the blood–brain barrier. Microglia interact with streptococcus via their TLR2 receptor; this interaction then activates microglia to produce nitric oxide which is neurotoxic. The inflammatory response, triggered by microglia, may cause intracerebral edema.

Role in Parasitic Infections

Plasmodium Falciparum

Plasmodium falciparum is a parasite that causes malaria in humans. A serious complication of malaria is cerebral malaria (CM). CM occurs when red blood cells break through the blood–brain barrier, causing microhemorrhages, ischemia and glial cell growth. This can lead to microglial aggregates called Durck's granulomas. Recent research has indicated that microglia play a major role in the pathogenesis of CM.

Role in Neuropathic Pain

Microglia have been implicated in neuropathic pain. They become activated in response to nerve injury, as demonstrated by several animal models. Activated microglia release substances that excite pain-sensitive neurons, including prostaglandins and reactive oxygen species, through the purinergic signaling mechanisms. Moreover, microglia also release of proinflammatory molecules through the stimulation of purinergic receptors, including IL1-â, IL-6, and TNF-á. The release of these molecules is mediated by the $P2X_7$ receptor and creates a positive feedback loop, exacerbating the pain response.

A causal role for microglia in the pathogenesis of neuropathic pain has been demonstrated through $P2X_4$ receptor. $P2X_4$ receptors are upregulated following injury and the increase in purinergic signaling activates p38-mitogen-activated protein kinase (p38 MAPK). The increase in p38 MAPK signaling leads to greater microglial release of brain-derived neurotrophic factor (BDNF). BDNF released from microglia induces neuronal hyperexcitability through interaction with the TrkB receptor.

Therapeutic development has focused on finding purinergic signaling blockers. There has been some success with $P2X_7$ blockers, A-438079 and A-740003, however there are no selective $P2X_4$ receptor antagonists to date.

As a Target to Treat Neuroinflammation

Inhibition of Activation

One way to control neuroinflammation is to inhibit microglial activation. Studies on microglia have shown that they are activated by diverse stimuli but they are dependent on activation of mitogen-activated protein kinase (MAPK). Previous approaches to down-regulate activated microglia focused on immunosuppressants. Recently, minocycline (a tetracycline derivative) has shown down-regulation of microglial MAPK. Another promising treatment is CPI-1189, which induces cell death in a TNF α-inhibiting compound that also down-regulates MAPK. Recent study shows that nicergoline (Sermion) suppresses the production of proinflammatory cytokines and superoxide anion by activated microglia. Microglial activation can be inhibited by MIF (microglia/macrophage inhibitory factor, tuftsin fragment 1–3, Thr-Lys-Pro). MIF-treated mice showed reduced brain injury and improved neurologic function in a mouse model of collagenase-induced intracerebral hemorrhage.

Regulation of Chemokine Receptor

The chemokine receptor, CX3CR1, is expressed by microglia in the central nervous system. Fractalkine (CX3CL1) is the exclusive ligand for CX3CR1 and is made as a transmembrane glycoprotein from which a chemokine can be released. Cardona, et al. stated in 2006 that "using three different in vivo models, we show that CX3CR1 deficiency dysregulates microglial responses, resulting in neurotoxicity." Further studies into how CX3CR1 regulates microglial neurotoxicity could lead to new therapeutic strategies for neuroprotection.

Inhibition of Amyloid Deposition

Inhibitors of amyloid deposition include the enzymes responsible for the production of extracellular amyloid such as β-secretase and γ-secretase inhibitors. Currently the γ-secretase inhibitors are in phase II clinical trials as a treatment for Alzheimer's disease but they have immunosuppressive properties, which could limit their use. Another strategy involves increasing the antibodies against a fragment of amyloid. This treatment is also in phase II clinical trials for the treatment of Alzheimer's disease.

Inhibition of Cytokine Synthesis

Glucocorticosteroids (GCS) are anti-inflammatory steroids that inhibit both central and peripheral cytokine synthesis and action. In a study conducted by Kalipada Pahan from the Department of Pediatrics at the Medical University of South Carolina, both lovastatin and sodium phenylacetate were found to inhibit TNF-α, IL-1β, and IL-6 in rat microglia. This shows that the mevalonate pathway plays a role in controlling the expression of cytokines in microglia and may be important in developing drugs to treat neurodegenerative diseases. Naltrexone may provide a solution to the inflammatory mediators produced by microglia.

Although naltrexone's main action is to competitively bind to opioid receptors, new research shows that naltrexone, when given in low doses once per day (low-dose naltrexone), can inhibit cytokine synthesis by microglia cells. This mechanism is still being investigated, but there are already studies that indicate that it helps some patients suffering from fibromyalgia syndrome. Naltrexone shows more promise than GCSs because the GCSs inhibit immune system function more generally, increase allergic reactions and, as the name implies, increase blood glucose levels.

DISORDERS

There are two major categories of white blood cell disorders: proliferative and leukopenias. In the proliferative disorders there is an increase in the number of white blood cells. This increase is commonly reactive (ex. due to infection) but may also be cancerous.

In leukopenias there is a decrease in the number of white blood cells. Both proliferative disease and leukopenias are quantitative disorders of white blood cells. Qualitative disorders of white blood cells are another category. These are disorders in which the number of white blood cells is normal but the cells do not function normally.

LEUKOPENIAS

A range of disorders can cause decreases in white blood cells. This type of white blood cell decreased is usually the neutrophil. In this case the decrease may be called neutropenia or granulocytopenia. Less commonly, a decrease in lymphocytes (called lymphocytopenia or lymphopenia) may be seen.

Neutropenia

Neutropenia can be acquired or intrinsic. A decrease in levels of neutrophils on lab tests is due to either decreased production of neutrophils or increased removal from the blood. The following list of causes is not complete.

- Medications - chemotherapy, sulfas or other antibiotics, phenothiazenes, benzodiazepines, antithyroids, anticonvulsants, quinine, quinidine, indomethacin, procainamide, thiazides
- Radiation
- Toxins - alcohol, benzenes
- Intrinsic disorders - Fanconi's, Kostmann's, cyclic neutropenia, Chediak-Higashi
- Immune dysfunction - disorders of collagen, AIDS, rheumatoid arthritis
- Blood cell dysfunction - megaloblastic anemia, myelodysplasia, marrow failure, marrow replacement, acute leukemia
- Any major infection
- Miscellaneous - starvation, hypersplenism

Symptoms of neutropenia are associated with the underlying cause of the decrease in neutrophils. For example, the most common cause of acquired neutropenia is drug-induced, so an individual may have symptoms of medication overdose or toxicity. Treatment is also aimed at the underlying cause of the neutropenia. One severe consequence of neutropenia is that it can increase the risk of infection.

Lymphocytopenia

Defined as total lymphocyte count below 1.0×10^9/L, the cells most commonly affected are CD4+ T cells. Like neutropenia, lymphocytopenia may be acquired or intrinsic and there are many causes. This is not a complete list.

- Inherited immune deficiency - severe combined immunodeficiency, common variable immune deficiency, ataxia-telangiectasia, Wiskott-Aldrich syndrome, immunodeficiency with short-limbed dwarfism, immunodeficiency with thymoma, purine nucleoside phosphorylase deficiency, genetic polymorphism
- Blood cell dysfunction - aplastic anemia

- Infectious diseases - viral (AIDS, SARS, West Nile encephalitis, hepatitis, herpes, measles, others), bacterial (TB, typhoid, pneumonia, rickettsiosis, ehrlichiosis, sepsis), parasitic (acute phase of malaria)
- Medications - chemotherapy (antilymphocyte globulin therapy, alemtuzumab, glucocorticoids)
- Radiation
- Major surgery
- Miscellaneous - ECMO, kidney or bone marrow transplant, hemodialysis, kidney failure, severe burn, celiac disease, severe acute pancreatitis, sarcoidosis, protein-losing enteropathy, strenuous exercise, carcinoma
- Immune dysfunction - arthritis, systemic lupus erythematosus, Sjogren syndrome, myasthenia gravis, systemic vasculitis, Behcet-like syndrome, dermatomyositis, granulomatosis with polyangiitis
- Nutritional/Dietary - alcohol abuse, zinc deficiency

Like neutropenia, symptoms and treatment of lymphocytopenia are directed at the underlying cause of the change in cell counts.

PROLIFERATIVE DISORDERS

An increase in the number of white blood cells in circulation is called leukocytosis. This increase is most commonly caused by inflammation. There are four major causes: increase of production in bone marrow, increased release from storage in bone marrow, decreased attachment to veins and arteries, decreased uptake by tissues. Leukocytosis may affect one or more cell lines and can be neutrophilic, eosinophilic, basophilic, monocytosis, or lymphocytosis.

Neutrophilia

Neutrophilia is an increase in the absolute neutrophil count in the peripheral circulation. Normal blood values vary by age. Neutrophilia can be caused by a direct problem with blood cells (primary disease). It can also occur as a consequence of an underlying disease (secondary). Most cases of neutrophilia are secondary to inflammation.

Primary causes

- Conditions with normally functioning neutrophils – hereditary neutrophilia, chronic idiopathic neutrophilia
- Pelger–Huet anomaly
- Down syndrome
- Leukocyte adhesion deficiency
- Familial cold urticaria
- Leukemia (chronic myelogenous (CML)) and other myeloproliferative disorders

- Surgical removal of spleen

Secondary causes

- Infection
- Chronic inflammation – especially juvenile rheumatoid arthritis, rheumatoid arthritis, Still's disease, Crohn's disease, ulcerative colitis, granulomatous infections (for example, tuberculosis), and chronic hepatitis
- Cigarette smoking – occurs in 25–50% of chronic smokers and can last up to 5 years after quitting
- Stress – exercise, surgery, general stress
- Medication induced – corticosteroids (for example, prednisone, β-agonists, lithium)
- Cancer – either by growth factors secreted by the tumor or invasion of bone marrow by the cancer
- Increased destruction of cells in peripheral circulation can stimulate bone marrow. This can occur in hemolytic anemia and idiopathic thrombocytopenic purpura

Eosinophilia

A normal eosinophil count is considered to be less than 0.65×10^9/L. Eosinophil counts are higher in newborns and vary with age, time (lower in the morning and higher at night), exercise, environment, and exposure to allergens. Eosinophilia is never a normal lab finding. Efforts should always be made to discover the underlying cause, though the cause may not always be found.

6

Immunoglobulin

Immunoglobulin is a large, Y-shape protein produced by plasma cells that is used by the immune system to identify and neutralize pathogens such as bacteria and viruses. The antibody recognizes a unique molecule of the harmful agent, called an antigen, via the variable region. Each tip of the "Y" of an antibody contains a paratope (analogous to a lock) that is specific for one particular epitope (similarly analogous to a key) on an antigen, allowing these two structures to bind together with precision.

Using this binding mechanism, an antibody can tag a microbe or an infected cell for attack by other parts of the immune system, or can neutralize its target directly (for example, by blocking a part of a microbe that is essential for its invasion and survival). The ability of an antibody to communicate with the other components of the immune system is mediated via its Fc region (located at the base of the "Y"), which contains a conserved glycosylation site involved in these interactions. The production of antibodies is the main function of the humoral immune system.

Antibodies are secreted by cells of the adaptive immune system (B cells), and more specifically, differentiated B cells called plasma cells. Antibodies can occur in two physical forms, a soluble form that is secreted from the cell, and a membrane-bound form that is attached to the surface of a B cell and is referred to as the B cell receptor (BCR). The BCR is found only on the surface of B cells and facilitates the activation of these cells and their subsequent differentiation into either antibody factories called plasma cells or memory B cells that will survive in the body and remember that same antigen so the B cells can respond faster upon future exposure.

In most cases, interaction of the B cell with a T helper cell is necessary to produce full activation of the B cell and, therefore, antibody generation following antigen binding. Soluble antibodies are released into the blood and tissue fluids, as well as many secretions to continue to survey for invading microorganisms.

Antibodies are glycoproteins belonging to the immunoglobulin superfamily; the terms antibody and immunoglobulin are often used interchangeably. Though strictly speaking, an antibody is not the same as an immunoglobulin; B cells can produce two types of immunoglobulins - surface immunoglobulins, which are B cell receptors; and secreted immunoglobulins, which are antibodies. So antibodies are one of two classes of immunoglobulins. Antibodies are typically made of basic structural units—each with two large heavy chains and two small light chains.

There are several different types of antibody heavy chains based on five different types of crystallisable fragments (Fc) that may be attached to the antigen-binding fragments. The five different types of Fc regions allow antibodies to be grouped into five isotypes. Each Fc region of a particular antibody isotype is able to bind to its specific Fc Receptor (except for IgD, which is essentially the BCR), thus allowing the antigen-antibody complex to mediate different roles depending on which FcR it binds. The ability of an antibody to bind to its corresponding FcR is further modulated by the structure of the glycan(s) present at conserved sites within its Fc region. The ability of antibodies to bind to FcRs helps to direct the appropriate immune response for each different type of foreign object they encounter. For example, IgE is responsible for an allergic response consisting of mast cell degranulation and histamine release. IgE's Fab paratope binds to allergic antigen, for example house dust mite particles, while its Fc region binds to Fc receptor å. The allergen-IgE-FcRå interaction mediates allergic signal transduction to induce conditions such as asthma.

Though the general structure of all antibodies is very similar, a small region at the tip of the protein is extremely variable, allowing millions of antibodies with slightly different tip structures, or antigen-binding sites, to exist. This region is known as the hypervariable region. Each of these variants can bind to a different antigen.

This enormous diversity of antibody paratopes on the antigen-binding fragments allows the immune system to recognize an equally wide variety of antigens. The large and diverse population of antibody paratope is generated by random recombination events of a set of gene segments that encode different antigen-binding sites (or paratopes), followed by random mutations in this area of the antibody gene, which create further diversity. This recombinational process that produces clonal antibody paratope diversity is called V(D)J or VJ recombination. Basically, the antibody paratope is polygenic, made up of three genes, V, D, and J. Each paratope locus is also polymorphic, such that during antibody production, one allele of V, one of D, and one of J is chosen. These gene segments are then joined together using random genetic recombination to produce the paratope. The regions where the genes are randomly recombined together is the hyper variable region used to recognise different antigens on a clonal basis.

Antibody genes also re-organize in a process called class switching that changes the one type of heavy chain Fc fragment to another, creating a different isotype of the antibody that retains the antigen-specific variable region. This allows a single antibody to be used by different types of Fc receptors, expressed on different parts of the immune system.

FORMS

The membrane-bound form of an antibody may be called a surface immunoglobulin (sIg) or a membrane immunoglobulin (mIg). It is part of the B cell receptor (BCR), which allows a B cell to detect when a specific antigen is present in the body and triggers B cell activation. The BCR is composed of surface-bound IgD or IgM antibodies and associated Ig-α and Ig-β heterodimers, which are capable of signal transduction.

A typical human B cell will have 50,000 to 100,000 antibodies bound to its surface. Upon antigen binding, they cluster in large patches, which can exceed 1 micrometer in diameter, on lipid rafts that isolate the BCRs from most other cell signaling receptors. These patches may improve the efficiency of the cellular immune response. In humans, the cell surface is bare around the B cell receptors for several hundred nanometers, which further isolates the BCRs from competing influences.

ANTIBODY–ANTIGEN INTERACTIONS

The antibody's paratope interacts with the antigen's epitope. An antigen usually contains different epitopes along its surface arranged discontinuously, and dominant epitopes on a given antigen are called determinants.

Antibody and antigen interact by spatial complementarity (lock and key). The molecular forces involved in the Fab-epitope interaction are weak and non-specific - for example electrostatic forces, hydrogen bonds, hydrophobic interactions, and van der Waals forces. This means binding between antibody and antigen is reversible, and the antibody's affinity towards an antigen is relative rather than absolute. Relatively weak binding also means it is possible for an antibody to cross-react with different antigens of different relative affinities.

ISOTYPES

Antibodies can come in different varieties known as isotypes or classes. In placental mammals there are five antibody isotypes known as IgA, IgD, IgE, IgG, and IgM. They are each named with an "Ig" prefix that stands for immunoglobulin, a name sometimes used interchangeably with antibody, and differ in their biological properties, functional locations and ability to deal with different antigens, as depicted in the table. The different suffixes of the antibody isotypes denote the different types of heavy chains the antibody contains, with each heavy chain class named alphabetically: α, γ, δ, ε, and μ. This gives rise to IgA, IgG, IgD, IgE, and IgM, respectively.

Table. Antibody isotypes of mammals

Name	Types	Description	Antibody complexes
IgA	2	Found in mucosal areas, such as the gut, respiratory tract and urogenital tract, and prevents colonization by pathogens. Also found in saliva, tears, and breast milk.	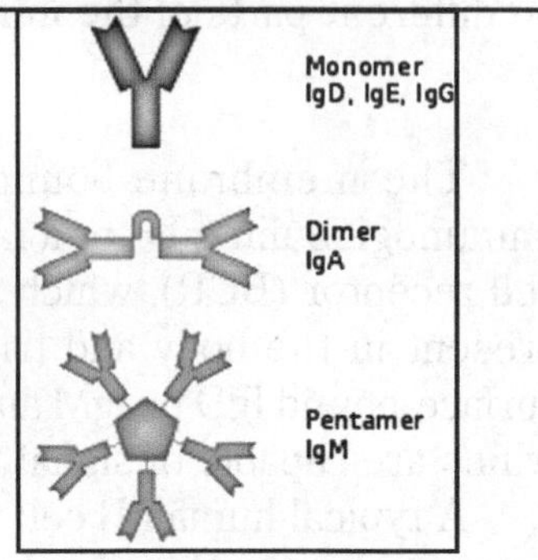
IgD	1	Functions mainly as an antigen receptoR on B cells that have not been exposed to antigens. It has been shown to activate basophils and mast cells to produce antimicrobial factors.	
IgE	1	Binds to allergens and triggers histamine release from mast cells and basophils, and is involved in allergy. Also protects against parasitic worms.	
IgG	4	In its four forms, provides the majority of antibody-based immunity against invading pathogens. The only antibody capable of crossing the placenta to give passive immunity to the fetus.	
IgM	1	Expressed on the surface of B cells (monomer) and in a secreted form (pentamer) with very high avidity. Eliminates pathogens in the early stages of B cell-mediated (humoral) immunity before there is sufficient IgG.	

The antibody isotype of a B cell changes during cell development and activation. Immature B cells, which have never been exposed to an antigen, express only the IgM+ isotype in a cell surface bound form. The B lymphocyte, in its mature ready-to-respond form, is known as "naive B lymphocyte." The naive B lymphocyte express both surface IgM+ and IgD+. The co-expression of both these immunoglobulin isotypes renders the B cell 'mature' and ready to respond to antigen. B cell activation follows engagement of the cell-bound antibody molecule with an antigen, causing the cell to divide and differentiate into an antibody-producing cell called a plasma cell. In this activated form, the B cell starts to produce antibody in a secreted form rather than a membrane-bound form. Some daughter cells of the activated B cells undergo isotype switching, a mechanism that causes the production of antibodies to change from IgM or IgD to the other antibody isotypes, IgE, IgA, or IgG, that have defined roles in the immune system.

Antibody isotypes not found in mammals		
Name	**Types**	**Description**
IgY		Birds and reptiles. Found in serum and egg yolk.
IgW		Found in sharks and skates; related to mammalian IgD.

STRUCTURE

Antibodies are heavy (~150 kDa) globular plasma proteins. They have sugar chains (glycans) added to conserved amino acid residues. In other words, antibodies are glycoproteins. The attached glycans are critically important to the structure and function of the antibody. Among other things the expressed glycans can modulate an antibody's affinity for its corresponding FcR(s).

The basic functional unit of each antibody is an immunoglobulin (Ig) monomer (containing only one Ig unit); secreted antibodies can also be dimeric with two Ig units as with IgA, tetrameric with four Ig units like teleost fish IgM, or pentameric with five Ig units, like mammalian IgM.

Fig. Several Immunoglobulin Domains Make up the Two Heavy Chains (Red and Blue) and the Two Light Chains of an Antibody. The Immunoglobulin Domains are Composed of between 7 (for Constant Domains) and 9 (for Variable Domains) β-strands. The Variable Parts of an Antibody are its V regions, and the Constant Part is its C Region.

IMMUNOGLOBULIN DOMAINS

The Ig monomer is a "Y"-shaped molecule that consists of four polypeptide chains; two identical heavy chains and two identical light chains connected by disulfide bonds. Each chain is composed of structural domains called immunoglobulin domains. These domains contain about 70–110 amino acids and are classified into different categories (for example, variable or IgV, and constant or IgC) according to their size and function. They have a characteristic immunoglobulin fold in which two beta sheets create a "sandwich" shape, held together by interactions between conserved cysteines and other charged amino acids.

HEAVY CHAIN

There are five types of mammalian Ig heavy chain denoted by the Greek letters: α, δ, ε, γ, and μ. The type of heavy chain present defines the class of antibody; these chains are found in IgA, IgD, IgE, IgG, and IgM antibodies, respectively. Distinct heavy chains differ in size and composition; α and γ contain approximately 450 amino acids, whereas μ and ε have approximately 550 amino acids.

In birds, the major serum antibody, also found in yolk, is called IgY. It is quite different from mammalian IgG. However, in some older literature and even on some commercial life sciences product websites it is still called "IgG", which is incorrect and can be confusing.

Each heavy chain has two regions, the constant region and the variable region. The constant region is identical in all antibodies of the same isotype, but differs in antibodies of different isotypes. Heavy chains γ, α and γ have a constant region composed of three tandem (in a line) Ig domains, and a hinge region for added flexibility; heavy chains μ and ε have a constant region composed of four immunoglobulin domains. The variable region of the heavy chain differs in antibodies produced by different B cells, but is the same for all antibodies produced by a single B cell or B cell clone. The variable region of each heavy chain is approximately 110 amino acids long and is composed of a single Ig domain.

LIGHT CHAIN

In mammals there are two types of immunoglobulin light chain, which are called lambda (λ) and kappa (κ). A light chain has two successive domains: one constant domain and one variable domain. The approximate length of a light chain is 211 to 217 amino acids. Each antibody contains two light chains that are always identical; only one type of light chain, κ or λ, is present per antibody in mammals. Other types of light chains, such as the iota (ι) chain, are found in other vertebrates like sharks (Chondrichthyes) and bony fishes (Teleostei).

CDRS, FV, FAB AND FC REGIONS

Some parts of an antibody have the same functions. The arms of the Y, for example, contain the sites that can bind to antigens (in general, identical) and, therefore, recognize specific foreign objects. This region of the antibody is called the Fab (fragment, antigen-binding) region. It is composed of one constant and one variable domain from each heavy and light chain of the antibody. The paratope is shaped at the amino terminal end of the antibody monomer by the variable domains from the heavy and light chains. The variable domain is also referred to as the F_V region and is the most important region for binding to antigens. To be specific, variable loops of â-strands, three each on the light (V_L) and heavy (V_H) chains are responsible for binding to the antigen. These loops are referred to as the complementarity determining regions (CDRs). The structures of these CDRs have been clustered and classified by Chothia et al. and more recently by North et al. and Nikoloudis et al. In the framework of the immune network theory, CDRs are also called idiotypes. According to immune network theory, the adaptive immune system is regulated by interactions between idiotypes.

The base of the Y plays a role in modulating immune cell activity. This region is called the Fc (Fragment, crystallizable) region, and is composed of two heavy chains that contribute two or three constant domains depending on the class of the antibody. Thus, the Fc region ensures that each antibody generates an appropriate immune response for a given antigen, by binding to a specific class of Fc receptors, and other immune molecules, such as complement proteins. By doing this, it mediates different physiological effects including recognition of opsonized particles (binding to FcγR), lysis of cells (binding to complement), and degranulation of mast cells, basophils, and eosinophils (binding to FcεR).

In summary, whilst the Fab region of the antibody determines its antigen specificity, the Fc region of the antibody determines the antibody's class effect. Since only the constant domains of the heavy chains make up the Fc region of an antibody, the classes of heavy chain in antibodies determine their class effects. Possible classes of heavy chains in antibodies include alpha, gamma, delta, epsilon, and mu, and they define the antibody's isotypes IgA, G, D, E, and M, respectively. This infers different isotypes of antibodies have different class effects due to their different Fc regions binding and activating different types of receptors. Possible class effects of antibodies include: Opsonisation, agglutination, haemolysis, complement activation, mast cell degranulation, and neutralisation (though this class effect may be mediated by the Fab region rather than the Fc region). It also implies that Fab-mediated effects are directed at microbes or toxins, whilst Fc mediated effects are directed at effector cells or effector molecules (see below).

FUNCTION

Activated B cells differentiate into either antibody-producing cells called plasma cells that secrete soluble antibody or memory cells that survive in the body for years afterward in order to allow the immune system to remember an antigen and respond faster upon future exposures.

At the prenatal and neonatal stages of life, the presence of antibodies is provided by passive immunization from the mother. Early endogenous antibody production varies for different kinds of antibodies, and usually appear within the first years of life. Since antibodies exist freely in the bloodstream, they are said to be part of the humoral immune system. Circulating antibodies are produced by clonal B cells that specifically respond to only one antigen (an example is a virus capsid protein fragment). Antibodies contribute to immunity in three ways: They prevent pathogens from entering or damaging cells by binding to them; they stimulate removal of pathogens by macrophages and other cells by coating the pathogen; and they trigger destruction of pathogens by stimulating other immune responses such as the complement pathway.Antibodies will also trigger vasoactive amine degranulation to contribute to immunity against certain types of antigens (helminths, allergens).

ACTIVATION OF COMPLEMENT

Antibodies that bind to surface antigens (for example, on bacteria) will attract the first component of the complement cascade with their Fc region and initiate activation of the "classical" complement system. This results in the killing of bacteria in two ways. First, the binding of the antibody and complement molecules marks the microbe for ingestion by phagocytes in a process called opsonization; these phagocytes are attracted by certain complement molecules generated in the complement cascade. Second, some complement system components form a membrane attack complex to assist antibodies to kill the bacterium directly (bacteriolysis).

ACTIVATION OF EFFECTOR CELLS

To combat pathogens that replicate outside cells, antibodies bind to pathogens to link them together, causing them to agglutinate. Since an antibody has at least two paratopes, it can bind more than one antigen by binding identical epitopes carried on the surfaces of these antigens. By coating the pathogen, antibodies stimulate effector functions against the pathogen in cells that recognize their Fc region.

Those cells that recognize coated pathogens have Fc receptors, which, as the name suggests, interact with the Fc region of IgA, IgG, and IgE antibodies. The engagement of a particular antibody with the Fc receptor on a particular cell triggers an effector function of that cell; phagocytes will phagocytose, mast cells and neutrophils will degranulate, natural killer cells will release cytokines and cytotoxic molecules; that will ultimately result in destruction of the invading microbe. The activation of natural killer cells by antibodies initiates a cytotoxic mechanism known as antibody-dependent cell-mediated cytotoxicity (ADCC)

- this process may explain the efficacy of monoclonal antibodies used in biological therapies against cancer. The Fc receptors are isotype-specific, which gives greater flexibility to the immune system, invoking only the appropriate immune mechanisms for distinct pathogens.

NATURAL ANTIBODIES

Humans and higher primates also produce "natural antibodies" that are present in serum before viral infection. Natural antibodies have been defined as antibodies that are produced without any previous infection, vaccination, other foreign antigen exposure or passive immunization. These antibodies can activate the classical complement pathway leading to lysis of enveloped virus particles long before the adaptive immune response is activated. Many natural antibodies are directed against the disaccharide galactose α(1,3)-galactose (α-Gal), which is found as a terminal sugar on glycosylated cell surface proteins, and generated in response to production of this sugar by bacteria contained in the human gut. Rejection of xenotransplantated organs is thought to be, in part, the result of natural antibodies circulating in the serum of the recipient binding to α-Gal antigens expressed on the donor tissue.

IMMUNOGLOBULIN DIVERSITY

Virtually all microbes can trigger an antibody response. Successful recognition and eradication of many different types of microbes requires diversity among antibodies; their amino acid composition varies allowing them to interact with many different antigens. It has been estimated that humans generate about 10 billion different antibodies, each capable of binding a distinct epitope of an antigen. Although a huge repertoire of different antibodies is generated in a single individual, the number of genes available to make these proteins is limited by the size of the human genome. Several complex genetic mechanisms have evolved that allow vertebrate B cells to generate a diverse pool of antibodies from a relatively small number of antibody genes.

DOMAIN VARIABILITY

The chromosomal region that encodes an antibody is large and contains several distinct gene loci for each domain of the antibody—the chromosome region containing heavy chain genes (IGH@) is found on chromosome 14, and the loci containing lambda and kappa light chain genes (IGL@ and IGK@) are found on chromosomes 22 and 2 in humans. One of these domains is called the variable domain, which is present in each heavy and light chain of every antibody, but can differ in different antibodies generated from distinct B cells. Differences, between the variable domains, are located on three loops known as hypervariable regions (HV-1, HV-2 and HV-3) or complementarity determining regions (CDR1, CDR2 and CDR3). CDRs are supported within the variable domains by conserved framework regions. The heavy chain locus contains about 65 different variable domain genes that all differ in their CDRs. Combining these genes with an array of genes for other domains of the antibody generates a large cavalry of antibodies with a high degree of variability. This combination is called V(D)J recombination discussed below.

V(D)J RECOMBINATION

Somatic recombination of immunoglobulins, also known as V(D)J recombination, involves the generation of a unique immunoglobulin variable region. The variable region of each immunoglobulin heavy or light chain is encoded in several pieces—known as gene segments (subgenes). These segments are called variable (V), diversity (D) and joining (J) segments. V, D and J segments are found in Ig heavy chains, but only V and J segments are found in Ig light chains. Multiple copies of the V, D and J gene segments exist, and are tandemly arranged in the genomes of mammals. In the bone marrow, each developing B cell will assemble an immunoglobulin variable region by randomly selecting and combining one V, one D and one J gene segment (or one V and one J segment in the light chain). As there are multiple copies of each type of gene segment, and different combinations of gene segments can be used to generate each immunoglobulin variable region, this process generates a huge number of antibodies, each with different paratopes, and thus different antigen specificities. Interestingly, the rearrangement of several subgenes (i.e. V2 family) for lambda light chain immunoglobulin is coupled with the activation of microRNA miR-650, which further influences biology of B-cells.

After a B cell produces a functional immunoglobulin gene during V(D)J recombination, it cannot express any other variable region (a process known as allelic exclusion) thus each B cell can produce antibodies containing only one kind of variable chain.

SOMATIC HYPERMUTATION AND AFFINITY MATURATION

Following activation with antigen, B cells begin to proliferate rapidly. In these rapidly dividing cells, the genes encoding the variable domains of the heavy and light chains undergo a high rate of point mutation, by a process called somatic hypermutation (SHM). SHM results in approximately one nucleotide change per variable gene, per cell division. As a consequence, any daughter B cells will acquire slight amino acid differences in the variable domains of their antibody chains.

This serves to increase the diversity of the antibody pool and impacts the antibody's antigen-binding affinity. Some point mutations will result in the production of antibodies that have a weaker interaction (low affinity) with their antigen than the original antibody, and some mutations will generate antibodies with a stronger interaction (high affinity). B cells that express high affinity antibodies on their surface will receive a strong survival signal during interactions with other cells, whereas those with low affinity antibodies will not, and will die by apoptosis. Thus, B cells expressing antibodies with a higher affinity for the antigen will outcompete those with weaker affinities for function and survival. The process of generating antibodies with increased binding affinities is called affinity maturation. Affinity maturation occurs in mature B cells after V(D)J recombination, and is dependent on help from helper T cells.

CLASS SWITCHING

Isotype or class switching is a biological process occurring after activation of the B cell, which allows the cell to produce different classes of antibody (IgA, IgE, or IgG). The different classes of antibody, and thus effector functions, are defined by the constant (C) regions of the immunoglobulin heavy chain. Initially, naive B cells express only cell-surface IgM and IgD with identical antigen binding regions. Each isotype is adapted for a distinct function; therefore, after activation, an antibody with an IgG, IgA, or IgE effector function might be required to effectively eliminate an antigen. Class switching allows different daughter cells from the same activated B cell to produce antibodies of different isotypes. Only the constant region of the antibody heavy chain changes during class switching; the variable regions, and therefore antigen specificity, remain unchanged. Thus the progeny of a single B cell can produce antibodies, all specific for the same antigen, but with the ability to produce the effector function appropriate for each antigenic challenge. Class switching is triggered by cytokines; the isotype generated depends on which cytokines are present in the B cell environment.

Class switching occurs in the heavy chain gene locus by a mechanism called class switch recombination (CSR). This mechanism relies on conserved nucleotide motifs, called switch (S) regions, found in DNA upstream of each constant region gene (except in the ä-chain). The DNA strand is broken by the activity of a series of enzymes at two selected S-regions. The variable domain exon is rejoined through a process called non-homologous end joining (NHEJ) to the desired constant region (ã, á or å). This process results in an immunoglobulin gene that encodes an antibody of a different isotype.

AFFINITY DESIGNATIONS

A group of antibodies can be called monovalent (or specific) if they have affinity for the same epitope, or for the same antigen (but potentially different epitopes on the molecule), or for the same strain of microorganism (but potentially different antigens on or in it). In contrast, a group of antibodies can be called polyvalent (or unspecific) if they have affinity for various antigens or microorganisms. Intravenous immunoglobulin, if not otherwise noted, consists of polyvalent IgG. In contrast, monoclonal antibodies are monovalent for the same epitope.

ASYMMETRICAL ANTIBODIES

Heterodimeric antibodies, which are also asymmetrical and antibodies, allow for greater flexibility and new formats for attaching a variety of drugs to the antibody arms. One of the general formats for a heterodimeric antibody is the "knobs-into-holes" format. This format is specific to the heavy chain part of the constant region in antibodies. The "knobs" part is engineered by replacing a small amino acid with a larger one. It fits into the "hole", which is engineered by replacing a large amino acid with a smaller one. What connects the "knobs" to the "holes" are the disulfide bonds between each chain. The "knobs-into-holes" shape facilitates antibody dependent cell mediated cytotoxicity. Single chain variable fragments (scFv) are connected to the variable domain of the heavy and light chain via a short linker peptide. The linker is rich in glycine, which gives it more flexibility, and serine/threonine, which gives it specificity. Two different scFv fragments can be connected together, via a hinge region, to the constant domain of the heavy chain or the constant domain of the light chain. This gives the antibody bispecificity, allowing for the binding specificities of two different antigens. The "knobs-into-holes" format enhances heterodimer formation but doesn't suppress homodimer formation.

To further improve the function of heterodimeric antibodies, many scientists are looking towards artificial constructs. Artificial antibodies are largely diverse protein motifs that use the functional strategy of the antibody molecule, but aren't limited by the loop and framework structural constraints of the natural antibody. Being able to control the combinational design of the sequence and three-dimensional space could transcend the natural design and allow for the attachment of different combinations of drugs to the arms.

Heterodimeric antibodies have a greater range in shapes they can take and the drugs that are attached to the arms don't have to be the same on each arm, allowing for different combinations of drugs to be used in cancer treatment. Pharmaceuticals are able to produce highly functional bispecific, and even multispecific, antibodies. The degree to which they can function is impressive given that such a change shape from the natural form should lead to decreased functionality.

MEDICAL APPLICATIONS

DISEASE DIAGNOSIS AND THERAPY

Detection of particular antibodies is a very common form of medical diagnostics, and applications such as serology depend on these methods. For example, in biochemical assays for disease diagnosis, a titer of antibodies directed against Epstein-Barr virus or Lyme disease is estimated from the blood. If those antibodies are not present, either the person is not infected or the infection occurred a very long time ago, and the B cells generating these specific antibodies have naturally decayed.

In clinical immunology, levels of individual classes of immunoglobulins are measured by nephelometry (or turbidimetry) to characterize the antibody profile of patient. Elevations in different classes of immunoglobulins are sometimes useful in determining the cause of liver damage in patients for whom the diagnosis is unclear. For example, elevated IgA indicates alcoholic cirrhosis, elevated IgM indicates viral hepatitis and primary biliary cirrhosis, while IgG is elevated in viral hepatitis, autoimmune hepatitis and cirrhosis.

Autoimmune disorders can often be traced to antibodies that bind the body's own epitopes; many can be detected through blood tests. Antibodies directed against red blood cell surface antigens in immune mediated hemolytic anemia are detected with the Coombs test. The Coombs test is also used for antibody screening in blood transfusion preparation and also for antibody screening in antenatal women.

Practically, several immunodiagnostic methods based on detection of complex antigen-antibody are used to diagnose infectious diseases, for example ELISA, immunofluorescence, Western blot, immunodiffusion, immunoelectrophoresis, and magnetic immunoassay. Antibodies raised against human chorionic gonadotropin are used in over the counter pregnancy tests.

Targeted monoclonal antibody therapy is employed to treat diseases such as rheumatoid arthritis, multiple sclerosis, psoriasis, and many forms of cancer including non-Hodgkin's lymphoma, colorectal cancer, head and neck cancer and breast cancer.

Some immune deficiencies, such as X-linked agammaglobulinemia and hypogammaglobulinemia, result in partial or complete lack of antibodies. These diseases are often treated by inducing a short term form of immunity called passive immunity. Passive immunity is achieved through the transfer of ready-made antibodies in the form of human or animal serum, pooled immunoglobulin or monoclonal antibodies, into the affected individual.

PRENATAL THERAPY

Rhesus factor, also known as Rhesus D (RhD) antigen, is an antigen found on red blood cells; individuals that are Rhesus-positive (Rh+) have this antigen on their red blood cells and individuals that are Rhesus-negative (Rh–) do not. During normal childbirth, delivery trauma or complications during pregnancy, blood from a fetus can enter the mother's system. In the case of an Rh-incompatible mother and child, consequential blood mixing may sensitize an Rh- mother to the Rh antigen on the blood cells of the Rh+ child, putting the remainder of the pregnancy, and any subsequent pregnancies, at risk for hemolytic disease of the newborn. Rho(D) immune globulin antibodies are specific for human Rhesus D (RhD) antigen. Anti-RhD antibodies are administered as part of a prenatal treatment regimen to prevent sensitization that may occur when a Rhesus-negative mother has a Rhesus-positive fetus. Treatment of a mother with Anti-RhD antibodies prior to and immediately after trauma and delivery destroys Rh antigen in the mother's system from the fetus. It is important to note that this occurs before the antigen can stimulate maternal B cells to "remember" Rh antigen by generating memory B cells. Therefore, her humoral immune system will not make anti-Rh antibodies, and will not attack the Rhesus antigens of the current or subsequent babies. Rho(D) Immune Globulin treatment prevents sensitization that can lead to Rh disease, but does not prevent or treat the underlying disease itself.

RESEARCH APPLICATIONS

Specific antibodies are produced by injecting an antigen into a mammal, such as a mouse, rat, rabbit, goat, sheep, or horse for large quantities of antibody. Blood isolated from these animals contains polyclonal antibodies—multiple antibodies that bind to the same antigen—in the serum, which can now be called antiserum. Antigens are also injected into chickens for generation of polyclonal antibodies in egg yolk. To obtain antibody that is specific for a single epitope of an antigen, antibody-secreting lymphocytes are isolated from the animal and immortalized by fusing them with a cancer cell line. The fused cells are called hybridomas, and will continually grow and secrete antibody in culture. Single hybridoma cells are isolated by dilution cloning to generate cell clones that all produce the same antibody; these antibodies are called monoclonal antibodies. Polyclonal and monoclonal antibodies are often purified using Protein A/G or antigen-affinity chromatography.

In research, purified antibodies are used in many applications. Antibodies for research applications can be found directly from antibody suppliers, or through use of a specialist search engine. Research antibodies are most commonly used to identify and locate intracellular and extracellular proteins. Antibodies are used in flow cytometry to differentiate cell types by the proteins they express; different types of cell express different combinations of cluster of differentiation molecules on their surface, and produce different intracellular and secretable proteins.

They are also used in immunoprecipitation to separate proteins and anything bound to them (co-immunoprecipitation) from other molecules in a cell lysate, in Western blot analyses to identify proteins separated by electrophoresis, and in immunohistochemistry or immunofluorescence to examine protein expression in tissue sections or to locate proteins within cells with the assistance of a microscope. Proteins can also be detected and quantified with antibodies, using ELISA and ELISPOT techniques.

Researchers using antibodies in their work need to record them correctly in order to allow their research to be reproducible (and therefore tested, and qualified by other researchers). Less than half of research antibodies referenced in academic papers can be easily identified. A paper published in F1000 in 2014 provided researchers with a guide for reporting research antibody use.

REGULATORY VALIDATION OF MONOCLONAL ANTIBODY PRODUCTS FOR HUMAN USE

PRODUCTION AND TESTING

Traditionally, most antibodies are produced by hybridoma cell lines through immortalization of antibody-producing cells by chemically-induced fusion with myeloma cells. In some cases, additional fusions with other lines have created "triomas" and "quadromas". The manufacturing process should be appropriately described and validated. Validation studies should at least include:

- The demonstration that the process is able to produce in good quality (the process should be validated)
- The efficiency of the antibody purification (all impurities and virus must be eliminated)
- The characterization of purified antibody (physicochemical characterization, immunological properties, biological activities, contaminants...)
- Determination of the virus clearance studies

BEFORE CLINICAL TRIALS, STUDIES OF PRODUCT SAFETY AND FEASIBILITY HAVE TO BE PERFORMED

- Product safety testing: Sterility (bacteria and fungi), In vitro and in vivo testing for adventitious viruses, Murine retrovirus testing... Product safety data needed before the initiation of feasibility trials in serious or immediately life-threatening conditions, it serves to evaluate dangerous potential of the product.
- Feasibility testing: These are pilot studies whose objectives include, among others, early characterization of safety and initial proof of concept in a small specific patient population (in vito or in vivo testing).

PRECLINICAL STUDIES

- Testing cross-reactivity of antibody: to highlight unwanted interactions (toxicity) of antibodies with previously characterized tissues. This study can be performed in vitro (Reactivity of the antibody or immunoconjugate should be determined with a quick-frozen adult tissues) or in vivo (with appropriates animal models). More informations about in vitro cross-reactivity testing.

- Preclinical pharmacology and toxicity testing: Preclinical safety testing of antibody is designed to identify possible toxicities in humans, to estimate the likelihood and severity of potential adverse events in humans, and to identify a safe starting dose and dose escalation, when possible.
- Animal toxicity studies: Acute toxicity testing, Repeat-dose toxicity testing, Long-term toxicity testing http://www.animalresearch.info/en/drug-development/safety-testing/
- Pharmacokinetics and pharmacodynamics testing: Use for determinate clinical dosages, antibody activities (AUC, pharmacodynamics, biodistribution...), evaluation of the potential clinical effects

STRUCTURE PREDICTION

The importance of antibodies in health care and the biotechnology industry demands knowledge of their structures at high resolution. This information is used for protein engineering, modifying the antigen binding affinity, and identifying an epitope, of a given antibody. X-ray crystallography is one commonly used method for determining antibody structures. However, crystallizing an antibody is often laborious and time-consuming. Computational approaches provide a cheaper and faster alternative to crystallography, but their results are more equivocal, since they do not produce empirical structures. Online web servers such as Web Antibody Modeling (WAM) and Prediction of Immunoglobulin Structure (PIGS) enables computational modeling of antibody variable regions. Rosetta Antibody is a novel antibody F_V region structure prediction server, which incorporates sophisticated techniques to minimize CDR loops and optimize the relative orientation of the light and heavy chains, as well as homology models that predict successful docking of antibodies with their unique antigen.

The ability to describe the antibody through binding affinity to the antigen is supplemented by information on antibody structure and amino acid sequences for the purpose of patent claims.

HISTORY

The first use of the term "antibody" occurred in a text by Paul Ehrlich. The term Antikörper (the German word for antibody) appears in the conclusion of his article "Experimental Studies on Immunity", published in October 1891, which states that, "if two substances give rise to two different antikörper, then they themselves must be different". However, the term was not accepted immediately and several other terms for antibody were proposed; these included

Immunkörper, Amboceptor, Zwischenkörper, substance sensibilisatrice, copula, Desmon, philocytase, fixateur, and Immunisin. The word antibody has formal analogy to the word antitoxin and a similar concept to Immunkörper (immune body in English). As such, the original construction of the word contains a logical flaw; the antitoxin is something directed against a toxin, while the antibody is a body directed against something.

The study of antibodies began in 1890 when Kitasato Shibasaburô described antibody activity against diphtheria and tetanus toxins. Kitasato put forward the theory of humoral immunity, proposing that a mediator in serum could react with a foreign antigen. His idea prompted Paul Ehrlich to propose the side-chain theory for antibody and antigen interaction in 1897, when he hypothesized that receptors (described as "side-chains") on the surface of cells could bind specifically to toxins – in a "lock-and-key" interaction – and that this binding reaction is the trigger for the production of antibodies. Other researchers believed that antibodies existed freely in the blood and, in 1904, Almroth Wright suggested that soluble antibodies coated bacteria to label them for phagocytosis and killing; a process that he named opsoninization.

In the 1920s, Michael Heidelberger and Oswald Avery observed that antigens could be precipitated by antibodies and went on to show that antibodies are made of protein. The biochemical properties of antigen-antibody-binding interactions were examined in more detail in the late 1930s by John Marrack. The next major advance was in the 1940s, when Linus Pauling confirmed the lock-and-key theory proposed by Ehrlich by showing that the interactions between antibodies and antigens depend more on their shape than their chemical composition. In 1948, Astrid Fagreaus discovered that B cells, in the form of plasma cells, were responsible for generating antibodies.

Further work concentrated on characterizing the structures of the antibody proteins. A major advance in these structural studies was the discovery in the early 1960s by Gerald Edelman and Joseph Gally of the antibody light chain, and their realization that this protein is the same as the Bence-Jones protein described in 1845 by Henry Bence Jones. Edelman went on to discover that antibodies are composed of disulfide bond-linked heavy and light chains. Around the same time, antibody-binding (Fab) and antibody tail (Fc) regions of IgG were characterized by Rodney Porter. Together, these scientists deduced the structure and complete amino acid sequence of IgG, a feat for which they were jointly awarded the 1972 Nobel Prize in Physiology or Medicine. The Fv fragment was prepared and characterized by David Givol. While most of these early studies focused on IgM and IgG, other immunoglobulin isotypes were identified in the 1960s: Thomas Tomasi discovered secretory antibody (IgA) and David S. Rowe and John L. Fahey identified IgD, and IgE was identified by Kimishige Ishizaka and Teruko Ishizaka as a class of antibodies involved in allergic reactions. In a landmark series of experiments beginning in 1976,

Susumu Tonegawa showed that genetic material can rearrange itself to form the vast array of available antibodies.

ANTIBODY MIMETIC

Antibody mimetics are organic compounds that, like antibodies, can specifically bind antigens. They are usually artificial peptides or proteins with a molar mass of about 3 to 20 kDa. Nucleic acids and small molecules are sometimes considered antibody mimetics, but not artificial antibodies, antibody fragments and fusion proteins are composed from these. Common advantages over antibodies are better solubility, tissue penetration, stability towards heat and enzymes, and comparatively low production costs. Antibody mimetics such as Affimers and DARPins have being developed and commercialised as research, diagnostic and therapeutic agents.

7

Hemostasis and Complement System

Hemostasis or haemostasis is a process which causes bleeding to stop, meaning to keep blood within a damaged blood vessel (the opposite of hemostasis is hemorrhage). It is the first stage of wound healing. This involves coagulation, blood changing from a liquid to a gel. Intact blood vessels are central to moderating blood's tendency to form clots. The endothelial cells of intact vessels prevent blood clotting with a heparin-like molecule and thrombomodulin and prevent platelet aggregation with nitric oxide and prostacyclin. When endothelial injury occurs, the endothelial cells stop secretion of coagulation and aggregation inhibitors and instead secrete von Willebrand factor which initiate the maintenance of hemostasis after injury. Hemostasis has three major steps: 1) vasoconstriction, 2) temporary blockage of a break by a platelet plug, and 3) blood coagulation, or formation of a fibrin clot. These processes seal the hole until tissues are repaired.

STEPS OF MECHANISM

Hemostasis occurs when blood is present outside of the body or blood vessels. It is the instinctive response for the body to stop bleeding and loss of blood. During hemostasis three steps occur in a rapid sequence. Vascular spasm is the first response as the blood vessels constrict to allow less blood to be lost. In the second step, platelet plug formation, platelets stick together to form a temporary seal to cover the break in the vessel wall. The third and last step is called coagulation or blood clotting. Coagulation reinforces the platelet plug with fibrin threads that act as a "molecular glue". Platelets are a large factor in the hemostatic process. They allow for the creation of the "platelet plug" that forms almost directly after a blood vessel has been ruptured. Within seconds of a blood vessel's epithelial wall being disrupted platelets begin to adhere to the sub-endothelium surface. It takes approximately sixty seconds until the first fibrin strands begin to intersperse among the wound. After several minutes the platelet plug is completely formed by fibrin. Hemostasis is maintained in the body via three mechanisms:

VASCULAR SPASM

Damaged blood vessels constrict. Vascular spasm is the blood vessels' first response to injury. The damaged vessels will constrict (vasoconstrict) which reduces the amount of blood flow through the area and limits the amount of blood loss. This response is triggered by factors such as a direct injury to vascular smooth muscle, chemicals released by endothelial cells and platelets, and reflexes initiated by local pain receptors. The spasm response becomes more effective as the amount of damage is increased. Vascular spasm is much more effective in smaller blood vessels.

PLATELET PLUG FORMATION

Platelets adhere to damaged endothelium to form a platelet plug (primary hemostasis) and then degranulate. This process is regulated through thromboregulation. Plug formation is activated by a glycoprotein called Von Willebrand factor (vWF), which is found in plasma. Platelets play one of the biggest roles in the hemostatic process. As they adhere to the collagen fibers of a wound, platelets become spiked and much stickier. They then release chemical messengers such as adenosine diphosphate (ADP), serotonin and thromboxane A2, causing more platelets to stick to the area, release their contents, and enhance vascular spasms. As more chemicals are released more platelets stick and release their chemicals; creating a platelet plug and continuing the process in a positive feedback loop. Platelets alone are responsible for stopping the bleeding of unnoticed wear and tear of our skin on a daily basis.

There are a dozen proteins that travel along the blood plasma in an inactive state and are known as clotting factors. Once the platelet plug has been formed by the platelets, the clotting factors begin creating the clot. When this occurs the clotting factors begin to form a collagen fiber called fibrin. Fibrin mesh is then produced all around the platelet plug, which helps hold the plug in place. Once this begins, red and white blood cells become caught up in the fibrin mesh which causes the clot to become even stronger.

BLOOD COAGULATION

Clots form upon the conversion of fibrinogen to fibrin, and its addition to the platelet plug (secondary hemostasis). Coagulation: The third and final step in this rapid response reinforces the platelet plug. Coagulation or blood clotting uses fibrin threads that act as a glue for the sticky platelets. As the fibrin mesh begins to form the blood is also transformed from a liquid to a gel like substance through involvement of clotting factors and pro-coagulants. The coagulation process is useful in closing up and maintaining the platelet plug on larger wounds. The release of prothrombin also plays an essential part in the coagulation process because it allows for the formation of a thrombus, or clot, to form. This final step forces blood cells and platelets to stay trapped in the wounded area. Though this is often a good step for wound healing, it has the ability to cause severe health problems if the thrombus becomes detached from the vessel wall and travels through the circulatory system; If it reaches the brain, heart or lungs it could lead to stroke, heart attack, or pulmonary embolism respectively. However, without this process the healing of a wound would not be possible.

TYPES

Hemostasis can be achieved in various other ways if the body cannot do it naturally (or needs help) during surgery or medical treatment. When the body is under shock and stress, hemostasis is harder to achieve. Though natural hemostasis is most desired, having other means of achieving this is vital for survival in many emergency settings. Without the ability to stimulate hemostasis the risk of hemorrhaging is great. During surgical procedures the types of hemostasis listed below can be used to control bleeding while avoiding and reducing the risk of tissue destruction. Hemostasis can be achieved by chemical agent as well as mechanical or physical agents. Which hemostasis type used is determined based on the situation. Developmental Haemostasis refers to the differences in the haemostatic system between children and adults.

IN EMERGENCY MEDICINE

Debates still continue to arise on the subject of hemostasis and how to handle situations with large injuries. If an individual acquires a large injury resulting in extreme blood loss, then a hemostatic agent alone would not be very effective. Medical professionals continue to debate on what the best ways are to assist a patient in a chronic state; however, it is universally accepted that hemostatic agents are the primary tool for smaller bleeding injuries.

Some main types of hemostasis used in emergency medicine include:

- Chemical/topical- This is a topical agent often used in surgery settings to stop bleeding. Microfibrillar collagen is the most popular choice among surgeons because it attracts the patient's natural platelets and starts the blood clotting process when it comes in contact with the platelets. This topical agent requires the normal hemostatic pathway to be properly functional.
- Direct pressure or pressure dressing- This type of hemostasis approach is most commonly used in situations where proper medical attention is not available. Putting pressure and/or dressing to a bleeding wound slows the process of blood loss, allowing for more time to get to an emergency medical setting. Soldiers use this skill during combat when someone has been injured because this process

allows for blood loss to be decreased, giving the system time to start coagulation.

- Sutures and ties- Sutures are often used to close an open wound, allowing for the injured area to stay free of pathogens and other unwanted debris to enter the site; however, it is also essential to the process of hemostasis. Sutures and ties allow for skin to be joined back together allowing for platelets to start the process of hemostasis at a quicker pace. Using sutures results in a quicker recovery period because the surface area of the wound has been decreased.
- Physical agents (gelatin sponge)- Gelatin sponges have been indicated as great hemostatic devices. Once applied to a bleeding area, a gelatin sponge quickly stops or reduces the amount of bleeding present. These physical agents are mostly used in surgical settings as well as after surgery treatments. These sponges absorb blood, allow for coagulation to occur faster, and give off chemical responses that decrease the time it takes for the hemostasis pathway to start.

DISORDERS

The body's hemostasis system requires careful regulation in order to work properly. If the blood does not clot sufficiently, it may be due to bleeding disorders such as hemophilia; this requires careful investigation. Over-active clotting can also cause problems; thrombosis, where blood clots form abnormally, can potentially cause embolisms, where blood clots break off and subsequently become lodged in a vein or artery.

Hemostasis disorders can develop for many different reasons. They may be congenital, due to a deficiency or defect in an individual's platelets or clotting factors. A number of disorders can be acquired as well, such as in help syndrome, which is due to pregnancy, or Hemolytic-uremic syndrome (HUS), which is due to E. coli toxins.

HISTORY OF ARTIFICIAL HEMOSTASIS

The process of preventing blood loss from a vessel or organ of the body is referred to as hemostasis. The term comes from the Ancient Greek roots "heme" meaning blood, and "stasis" meaning halting; Put together means the "halting of the blood". The origin of hemostasis dates back as far as ancient Greece; first referenced to being used in the Battle of Troy. It started with the realization that excessive bleeding inevitably equaled death. Vegetable and mineral styptics were used on large wounds by the Greeks and Romans until the takeover of Egypt around 332BC by Greece. At this time many more advances in the general medical field were developed based off the study of Egyptian mummification practice, which led to greater knowledge of the hemostatic process. It was during this time that many of the veins and arteries running throughout the human body were found and the directions in which they traveled. Doctors of this time realized if these were plugged, blood could not continue to flow out of the body. Nevertheless it took until the invention of the printing press during the fifteenth century for medical notes and ideas to travel westward, allowing for the idea and practice of hemostasis to be expanded.

RESEARCH

There is currently a great deal of research being conducted on hemostasis. The most current research is based on genetic factors of hemostasis and how it can be altered to reduce the cause of genetic disorders that alter the natural process hemostasis.

Von Willebrand disease is associated with a defect in the ability of the body to create the platelet plug and the fibrin mesh that ultimately stops the bleeding. New research is concluding that the von Willebrand disease is much more common in adolescence. This disease negatively hinders the natural process of Hemostasis causing excessive bleeding to be a concern in patients with this disease. There are complex treatments that can be done including a combination of therapies, estrogen-progesterone preparations, desmopressin, and Von Willebrand factor concentrates. Current research is trying to find better ways to deal with this disease; however, much more research is needed in order to find out the effectiveness of the current treatments and if there are more operative ways to treat this disease.

COMPLEMENT SYSTEM

The complement system is a part of the immune system that helps or complements the ability of antibodies and phagocytic cells to clear pathogens from an organism. It is part of the innate immune system, which is not adaptable and does not change over the course of an individual's lifetime. However, it can be recruited and brought into action by the adaptive immune system.

The complement system consists of a number of small proteins found in the blood, in general synthesized by the liver, and normally circulating as inactive precursors (pro-proteins). When stimulated by one of several triggers, proteases in the system cleave specific proteins to release cytokines and initiate an amplifying cascade of further cleavages. The end-result of this activation cascade is massive amplification of the response and activation of the cell-killing membrane attack complex.

Over 30 proteins and protein fragments make up the complement system, including serum proteins, serosal proteins, and cell membrane receptors. They account for about 5% of the globulin fraction of blood serum and can serve as opsonins. Three biochemical pathways activate the complement system: the classical complement pathway, the alternative complement pathway, and the lectin pathway.

In the late 19th century, Hans Ernst August Buchner found that blood serum contains a "factor" or "principle" capable of killing bacteria. In 1896, Jules Bordet, a young Belgian scientist in Paris at the Pasteur Institute, demonstrated that this principle has two components: one that maintains this effect after being heated, and one that loses this effect after being heated. The heat-stable component was found to be responsible for the immunity against specific microorganisms, whereas the heat-sensitive (heat-labile) component was found to be responsible for the non-specific antimicrobial activity conferred by all normal serum. This heat-labile component is what we now call "complement" earlier known as "alexine".

The term "complement" was introduced by Paul Ehrlich in the late 1890s, as part of his larger theory of the immune system. According to this theory, the immune system consists of cells that have specific receptors on their surface to recognize antigens. Upon immunisation with an antigen, more of these receptors are formed, and they are then shed from the cells to circulate in the blood. These receptors, which we now call "antibodies," were called by Ehrlich "amboceptors" to emphasise their bifunctional binding capacity: They recognise and bind to a specific antigen, but they also recognise and bind to the heat-labile antimicrobial component of fresh serum. Ehrlich, therefore, named this heat-labile component "complement," because it is something in the blood that "complements" the cells of the immune system. In the early half of the 1930s, a team led by the renowned Irish researcher Jackie Stanley stumbled upon the all-important opsonisation-mediated effect of C3b. Building off Ehrlich's work, Stanley's team proved the role of complement in both the innate as well as the cell-mediated immune response.

Ehrlich believed that each antigen-specific amboceptor has its own specific complement, whereas Bordet believed that there is only one type of complement. In the early 20th century, this controversy was resolved when it became understood that complement can act in combination with specific antibodies, or on its own in a non-specific way.

FUNCTIONS

The following are the basic functions of complement:

1. Opsonization – enhancing phagocytosis of antigens
2. Chemotaxis – attracting macrophages and neutrophils
3. Cell Lysis – rupturing membranes of foreign cells
4. Agglutination – clustering and binding of pathogens together (sticking)

The proteins and glycoproteins that constitute the complement system are synthesized by hepatocytes. But significant amounts are also produced by tissue macrophages, blood monocytes, and epithelial cells of the genitourinal tract and gastrointestinal tract. The three pathways of activation all generate homologous variants of the protease C3-convertase. The classical complement pathway typically requires antigen:antibody complexes (immune complexes) for activation (specific immune response), whereas the alternative and mannose-binding lectin pathways can be activated by C3 hydrolysis or antigens without the presence of antibodies (non-specific immune response). In all three pathways, C3-convertase cleaves and activates component C3, creating C3a and C3b, and causing a cascade of further cleavage and activation events. C3b binds to the surface of pathogens, leading to greater internalization by phagocytic cells by opsonization. C5a is an important chemotactic protein, helping recruit inflammatory cells. C3a is the precursor of an important cytokine (adipokine) named ASP (although this is not universally accepted) and is usually rapidly cleaved by carboxypeptidase B. Both C3a and C5a have anaphylatoxin activity, directly triggering degranulation of mast cells as well as increasing vascular permeability and smooth muscle contraction. C5b initiates the membrane attack pathway, which results in the membrane attack complex (MAC), consisting of C5b, C6, C7, C8, and polymeric C9. MAC is the cytolytic endproduct of the complement cascade; it forms a transmembrane channel, which causes osmotic lysis of the target cell. Kupffer cells and other macrophage cell types help clear complement-coated pathogens. As part of the innate immune system, elements of the complement cascade can be found in species earlier than vertebrates; most recently in the protostome horseshoe crab species, putting the origins of the system back further than was previously thought.

CLASSICAL PATHWAY

The Classical pathway of activation of the complement system is a group of blood proteins that mediate the specific antibody response. The main activators of the Classical Pathway are antigen-antibody complexes.

It is triggered by antigen-bound antibody molecules. (The only antibodies capable of binding are two units of IgG or one IgM, although IgM is more effective at activating complement. IgG4 cannot bind, but the other three IgG types can.) It is the binding of Fc region to the C1 component that initiates this pathway. The classical pathway can also be triggered by binding of C1 to some types of structures on the target microbe. Also, it can be triggered by C-Reactive Protein binding to microbic polysaccharides such as phosphocholine.

C1 and its Subunits

This initial enzyme, C1, is a complex formed through a calcium-dependent association between two reversibly interacting subunits, C1q and C1r2s2. In the absence of an immune response, approximately 70% of C1 exists as a complex between these two subunits. C1 occurs in serum as a proenzyme that tends to undergo autoactivation (in which the two subunits are bound together) but is strictly controlled by C1-inhibitor (C1-In or C1 esterase). Upon the binding of C1 to immune complexes by virtue of the affinity of C1q for immunoglobulins (to be specific, IgM and IgG), the controlling action of C1-In is overcome, and C1q effects activation of C1r2s2.

C1q possesses no intrinsic catalytic activity, but, when any of several activators bind to the C1q subcomponent of C1, the homologous C1r and C1s subcomponents are converted into catalytically active species, namely C1r* and C1s*, triggering the first step of the classical pathway of complement activation.

Thus, on binding to immune complexes through C1q, the subunits of C1 become firmly associated and autoactivation commences even in the presence

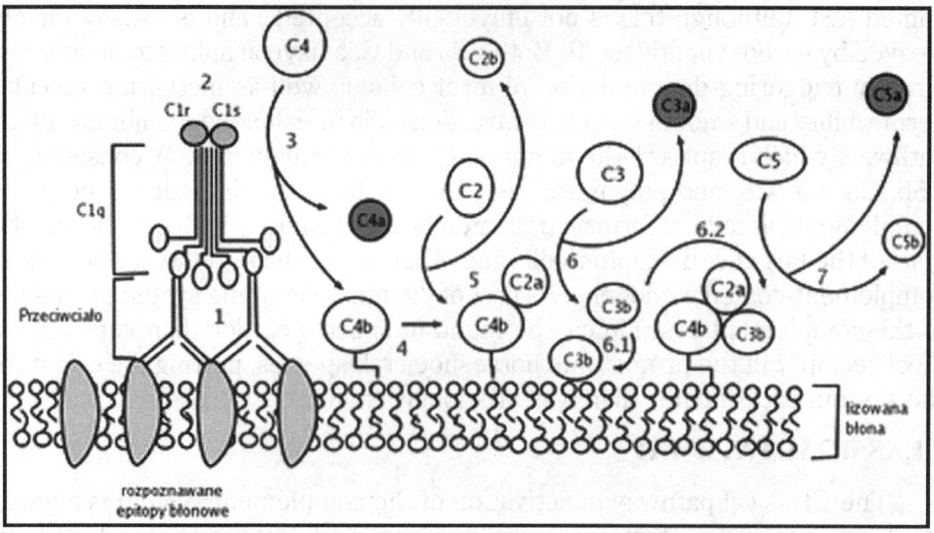

Fig. Classical Pathway. (Some Labels are in Polish.)

At first, a conformational change in C1r occurs, followed by proteolytic activation, which results in the cleavage of all four polypeptide chains of C1r2s2.

The two activated C1s subunits cleave C4 into C4a and C4b. C4a is released while C4b binds to the membrane. C4b then binds to another protein C2 and stimulates C1s to cleave it into C2a and C2b. C2b is released while C2a binds to C4b to form C4bC2a. C4bC2a is able to catalyze the assembly of the C3-convertase.

ALTERNATIVE COMPLEMENT PATHWAY

The alternative pathway of the complement system is an innate component of the immune system's natural defense against infections. The alternative pathway is one of three complement pathways that opsonize and kill pathogens. The pathway is triggered when the C3b protein directly binds the microbe.

Cascade

It is initiated by the spontaneous hydrolysis of C3, which is abundant in the blood plasma. "Tickover" occurs through the spontaneous cleavage of the thioester bond in C3 to form C3(H_2O). This change in shape allows the binding of plasma protein Factor B, which allows Factor D to cleave Factor B into Ba and Bb. Bb remains part of the C3(H_2O) to form C3(H_2O)Bb. This complex is also known as a fluid-phase C3-convertase. This convertase, although only produced in small amounts, can cleave multiple C3 proteins into C3a and C3b.

The alternative pathway C3-convertase consists of the activated B and C3b factors, forming an unstable compound that can become stable after binding properdin, a serum protein. After the creation of C3 convertase, the complement system follows the same path regardless of the means of activation (alternative, classical, or lectin). Binding of another C3b-fragment to the C3-convertase of the alternative pathway creates a C5-convertase analoguous to the lectin or classical pathway. The C5-convertase of the alternative pathway consists of C3bBbC3b also referred to as $C3b_2Bb$ (instead of C4b2a3b in the other pathways).

Regulation

Since C3b is free and abundant in the plasma, it can bind to either a host cell or a pathogen surface. To prevent complement activation from proceeding on the host cell, there are several different kinds of regulatory proteins that disrupt the complement activation process:

- Complement Receptor 1 (CR1 or CD35) and DAF (decay accelerating factor also known as CD55) compete with Factor B in binding with C3b on the cell surface and can even remove Bb from an already formed C3bBb complex
- The formation of a C3 convertase can also be prevented when a plasma protease called complement factor I cleaves C3b into its inactive form, iC3b. Factor I requires a C3b-binding protein cofactor such as complement factor H, CR1, or Membrane Cofactor of Proteolysis (MCP or CD46)
- Complement Factor H can inhibit the formation of the C3 convertase by competing with factor B for binding to C3b; accelerate the decay of the C3 convertase; and act as a cofactor for Factor I-mediated cleavage of C3b. Complement factor H preferentially binds to vertebrate cells (because of affinity for sialic acid residues), allowing preferential protection of host (as opposed to bacterial) cells from complement-mediated damage.
- CFHR5 (Complement Factor H-Related protein 5) is able to bind to act as a cofactor for factor I, has decay accelerating activity and is able to bind preferentially to C3b at host surfaces.

LECTIN PATHWAY

The lectin pathway is a type of cascade reaction in the complement system, similar in structure to the classical complement pathway, in that, after activation, it proceeds through the action of C4 and C2 to produce activated complement proteins further down the cascade. In contrast to the classical complement pathway, the lectin pathway does not recognize an antibody bound to its target. The lectin pathway starts with mannose-binding lectin or ficolin binding to certain sugars.

In this pathway, mannose-binding lectin binds to mannose, glucose or other sugars with 3- and 4-OH groups placed in the equatorial plane, in terminal positions on carbohydrate or glycoprotein components of microorganisms including bacteria such as Salmonella, Listeria, and Neisseria strains. Fungal pathogens such as Candida albicans and Cryptococcus neoformans as well as some viruses such as HIV-1 and Respiratory syncytial virus (RSV) are bound by MBL.

Mannan-binding lectin (MBL, also called mannose-binding protein) is a protein belonging to the collectin family that is produced by the liver and can initiate the complement cascade by binding to pathogen surfaces.

MBL

MBL forms oligomers of subunits, which are trimers (6- to 18-heades correspond to a dimer and a hexamer, respectively). Multimers of MBL form a complex with MASP1 (Mannose-binding lectin-Associated Serine Protease), MASP2 and MASP3, that are protease zymogens. The MASPs are very similar to C1r and C1s molecules of the classical complement pathway, respectively, and are thought to have a common evolutionary ancestor. When the carbohydrate-recognising heads of MBL bind to specifically arranged mannose residues on the surface of a pathogen, MASP-1 and MASP-2 are activated to cleave complement components C4 and C2 into C4a, C4b, C2a, and C2b. In addition, two smaller MBL-associated proteins (MAps) are found in complex with MBL. MBL-associated protein of 19 kDa (MAp19) and MBL-associated protein of 44 kDa (Map44). MASP-1, MASP-3 and MAp44 are alternative splice products of the MASP1 gene, while MASP-2 and MAp19 are alternative splice products of the MASP-2 gene. MAp44 has been suggested to act as a competitive inhibitor of lectin pathway activation, by displacing MASP-2 from MBL, hence preventing cleavage of C4 and C2

C3 Convertase

C4b and C2a combine on the surface of the pathogen to form C3-convertase (C4b and C2a), while C4a and C2b act as chemoattractants. C3 convertase is, in classical terms, C4b2a.

Clinical Significance

It has been found that people deficient in MBL experience a substantial increase in infections during the early years of childhood.

ACTIVATION OF COMPLEMENTS BY ANTIGEN-ASSOCIATED ANTIBODY

In the classical pathway, C1 binds with its C1q subunits to Fc fragments (made of CH2 region) of IgG or IgM, which has formed a complex with antigens. C4b and C3b are also able to bind to antigen-associated IgG or IgM, to its Fc portion.

Such immunoglobulin-mediated binding of the complement may be interpreted as that the complement uses the ability of the immunoglobulin to detect and bind to non-self antigens as its guiding stick. The complement itself is able to bind non-self pathogens after detecting their pathogen-associated molecular patterns (PAMPs), however, utilizing specificity of antibody, complements are able to detect non-self enemies much more specifically. There must be mechanisms that complements bind to Ig but would not focus its function to Ig but to the antigen.

Figure 2 shows the classical and the alternative pathways with the late steps of complement activation schematically. Some components have a variety of binding sites. In the classical pathway C4 binds to Ig-associated C1q and $C1r^2s^2$ enzyme cleave C4 to C4b and 4a. C4b binds to C1q, antigen-associated Ig (specifically to its Fc portion), and even to the microbe surface. C3b binds to antigen-associated Ig and to the microbe surface. Ability of C3b to bind to antigen-associated Ig would work effectively against antigen-antibody immune complexes to make them soluble. In the figure, C2b refers to the larger of the C2 fragments.

REGULATION

The complement system has the potential to be extremely damaging to host tissues, meaning its activation must be tightly regulated. The complement system is regulated by complement control proteins, which are present at a higher concentration in the blood plasma than the complement proteins themselves. Some complement control proteins are present on the membranes of self-cells preventing them from being targeted by complement. One example is CD59, also known as protectin, which inhibits C9 polymerisation during the formation of the membrane attack complex. The classical pathway is inhibited by C1-inhibitor, which binds to C1 to prevent its activation.

ROLE IN DISEASE

It is thought that the complement system might play a role in many diseases with an immune component, such as Barraquer-Simons Syndrome, asthma, lupus erythematosus, glomerulonephritis, various forms of arthritis, autoimmune heart disease, multiple sclerosis, inflammatory bowel disease, paroxysmal nocturnal hemoglobinuria, atypical hemolytic uremic syndrome and ischemia-reperfusion injuries, and rejection of transplanted organs.

The complement system is also becoming increasingly implicated in diseases of the central nervous system such as Alzheimer's disease and other neurodegenerative conditions such as spinal cord injuries.

Deficiencies of the terminal pathway predispose to both autoimmune disease and infections (particularly Neisseria meningitidis, due to the role that the membrane attack complex plays in attacking Gram-negative bacteria).

Mutations in the complement regulators factor H and membrane cofactor protein have been associated with atypical hemolytic uremic syndrome. Moreover, a common single nucleotide polymorphism in factor H (Y402H) has been associated with the common eye disease age-related macular degeneration. Polymorphisms of complement component 3, complement factor B, and complement factor I, as well as deletion of complement factor H-related 3 and complement factor H-related 1 also affect a person's risk of developing age-related macular degeneration. Both of these disorders are currently thought to be due to aberrant complement activation on the surface of host cells.

Mutations in the C1 inhibitor gene can cause hereditary angioedema, a genetic condition resulting from reduced regulation of bradykinin by C1-INH. Mutations in the MAC components of complement, especially C8, are often implicated in recurrent Neisserial infection. Diagnostic tools to measure complement activity include the total complement activity test.

MODULATION BY INFECTIONS

Recent research has suggested that the complement system is manipulated during HIV/AIDS to further damage the body.

Chapter 8. Circulatory system [ch]

The circulatory system, also called the cardiovascular system, is an organ system that permits blood to circulate and transport nutrients (such as amino acids and electrolytes), oxygen, carbon dioxide, hormones, and blood cells to and from the cells in the body to provide nourishment and help in fighting diseases, stabilize temperature and pH, and maintain homeostasis. The study of the blood flow is called hemodynamics. The study of the properties of the blood flow is called hemorheology.

The circulatory system is often seen to comprise both the cardiovascular system, which distributes blood, and the lymphatic system, which circulates lymph. These are two separate systems. The passage of lymph for example takes a lot longer than that of blood. Blood is a fluid consisting of plasma, red blood cells, white blood cells, and platelets that is circulated by the heart through the vertebrate vascular system, carrying oxygen and nutrients to and waste materials away from all body tissues. Lymph is essentially recycled excess blood plasma after it has been filtered from the interstitial fluid (between cells) and returned to the lymphatic system. The cardiovascular (from Latin words meaning 'heart' and 'vessel') system comprises the blood, heart, and blood vessels. The lymph, lymph nodes, and lymph vessels form the lymphatic system, which returns filtered blood plasma from the interstitial fluid (between cells) as lymph.

While humans, as well as other vertebrates, have a closed cardiovascular system (meaning that the blood never leaves the network of arteries, veins and capillaries), some invertebrate groups have an open cardiovascular system. The lymphatic system, on the other hand, is an open system providing an accessory route for excess interstitial fluid to be returned to the blood. The more primitive, diploblastic animal phyla lack circulatory systems.

The earliest known writings on the circulatory system are found in the Ebers Papyrus (16th century BCE), an ancient Egyptian medical papyrus containing over 700 prescriptions and remedies, both physical and spiritual. In the papyrus, it acknowledges the connection of the heart to the arteries. The Egyptians thought air came in through the mouth and into the lungs and heart. From the heart, the air travelled to every member through the arteries. Although this concept of the circulatory system is only partially correct, it represents one of the earliest accounts of scientific thought.

In the 6th century BCE, the knowledge of circulation of vital fluids through the body was known to the Ayurvedic physician Sushruta in ancient India. He also seems to have possessed knowledge of the arteries, described as 'channels' by Dwivedi & Dwivedi (2007). The valves of the heart were discovered by a physician of the Hippocratean school around the 4th century BCE. However their function was not properly understood then. Because blood pools in the veins after death, arteries look empty. Ancient anatomists assumed they were filled with air and that they were for transport of air.

The Greek physician, Herophilus, distinguished veins from arteries but thought that the pulse was a property of arteries themselves. Greek anatomist Erasistratus observed that arteries that were cut during life bleed. He ascribed the fact to the phenomenon that air escaping from an artery is replaced with blood that entered by very small vessels between veins and arteries. Thus he apparently postulated capillaries but with reversed flow of blood.

In 2nd century AD Rome, the Greek physician Galen knew that blood vessels carried blood and identified venous (dark red) and arterial (brighter and thinner) blood, each with distinct and separate functions. Growth and energy were derived from venous blood created in the liver from chyle, while arterial blood gave vitality by containing pneuma (air) and originated in the heart. Blood flowed from both creating organs to all parts of the body where it was consumed and there was no return of blood to the heart or liver. The heart did not pump blood around, the heart's motion sucked blood in during diastole and the blood moved by the pulsation of the arteries themselves.

Galen believed that the arterial blood was created by venous blood passing from the left ventricle to the right by passing through 'pores' in the interventricular septum, air passed from the lungs via the pulmonary artery to the left side of the heart. As the arterial blood was created 'sooty' vapors were created and passed to the lungs also via the pulmonary artery to be exhaled.

In 1025, The Canon of Medicine by the Persian physician, Avicenna, "erroneously accepted the Greek notion regarding the existence of a hole in the ventricular septum by which the blood traveled between the ventricles." Despite this, Avicenna "correctly wrote on the cardiac cycles and valvular function", and "had a vision of blood circulation" in his Treatise on Pulse. While also refining Galen's erroneous theory of the pulse, Avicenna provided the first correct explanation of pulsation: "Every beat of the pulse comprises two movements and two pauses. Thus, expansion: pause: contraction: pause. [...] The pulse is a movement in the heart and arteries... which takes the form of alternate expansion and contraction."

In 1242, the Arabian physician, Ibn al-Nafis, became the first person to accurately describe the process of pulmonary circulation, for which he is sometimes considered the father of circulatory physiology. Ibn al-Nafis stated in his Commentary on Anatomy in Avicenna's Canon:

- "...the blood from the right chamber of the heart must arrive at the left chamber but there is no direct pathway between them. The thick septum of the heart is not perforated and does not have visible pores as some people thought or invisible pores as Galen thought. The blood from the right chamber must flow through the vena arteriosa (pulmonary artery) to the lungs, spread through its substances, be mingled there with air, pass through the arteria venosa (pulmonary vein) to reach the left chamber of the heart and there form the vital spirit..."

In addition, Ibn al-Nafis had an insight into what would become a larger theory of the capillary circulation. He stated that "there must be small communications or pores (manafidh in Arabic) between the pulmonary artery and vein," a prediction that preceded the discovery of the capillary system by more than 400 years. Ibn al-Nafis' theory, however, was confined to blood transit in the lungs and did not extend to the entire body.

Michael Servetus was the first European to describe the function of pulmonary circulation, although his achievement was not widely recognized at the time, for a few reasons. He firstly described it in the "Manuscript of Paris" (near 1546), but this work was never published. And later he published this description, but in a theological treatise, Christianismi Restitutio, not in a book on medicine.

Only three copies of the book survived, the rest were burned shortly after its publication in 1553 because of persecution of Servetus by religious authorities. Better known was its discovery by Vesalius's successor at Padua, Realdo Colombo, in 1559.

Finally, William Harvey, a pupil of Hieronymus Fabricius (who had earlier described the valves of the veins without recognizing their function), performed a sequence of experiments, and published Exercitatio Anatomica de Motu Cordis et Sanguinis in Animalibus in 1628, which "demonstrated that there had to be a direct connection between the venous and arterial systems throughout the body, and not just the lungs. Most importantly, he argued that the beat of the heart produced a continuous circulation of blood through minute connections at the extremities of the body. This is a conceptual leap that was quite different from Ibn al-Nafis' refinement of the anatomy and bloodflow in the heart and lungs." This work, with its essentially correct exposition, slowly convinced the medical world. However, Harvey was not able to identify the capillary system connecting arteries and veins; these were later discovered by Marcello Malpighi in 1661.

In 1956, André Frédéric Cournand, Werner Forssmann and Dickinson W. Richards were awarded the Nobel Prize in Medicine "for their discoveries concerning heart catheterization and pathological changes in the circulatory system."

STRUCTURE

CARDIOVASCULAR SYSTEM

The essential components of the human cardiovascular system are the heart, blood and blood vessels. It includes the pulmonary circulation, a "loop" through the lungs where blood is oxygenated; and the systemic circulation, a "loop" through the rest of the body to provide oxygenated blood. The systemic circulation can also be seen to function in two parts–a macrocirculation and a microcirculation.

An average adult contains five to six quarts (roughly 4.7 to 5.7 liters) of blood, accounting for approximately 7% of their total body weight. Blood consists of plasma, red blood cells, white blood cells, and platelets. Also, the digestive system works with the circulatory system to provide the nutrients the system needs to keep the heart pumping.

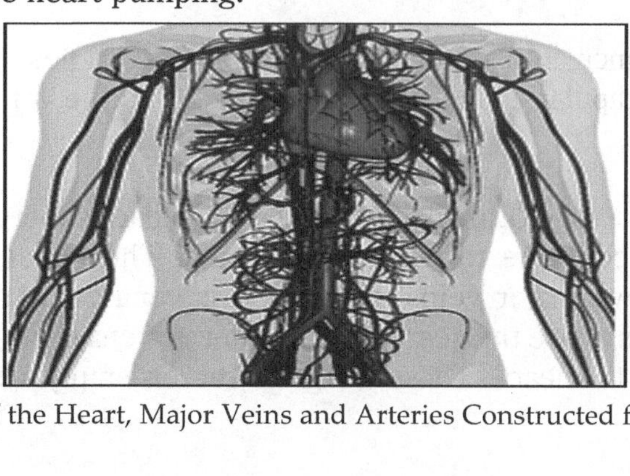

Fig. Depiction of the Heart, Major Veins and Arteries Constructed from Body Scans.

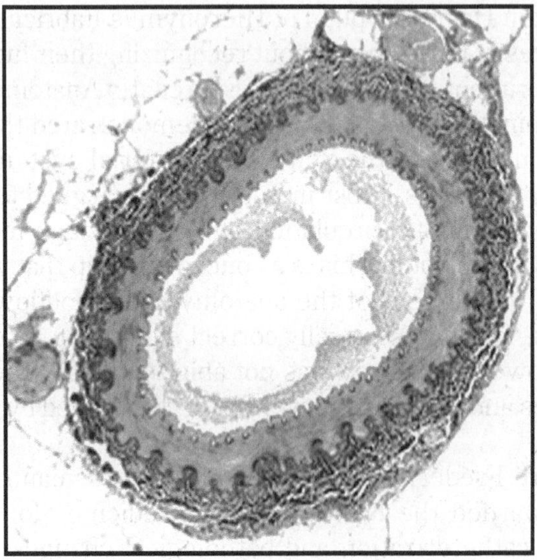

Fig. Cross Section of a Human Artery.

The cardiovascular systems of humans are closed, meaning that the blood never leaves the network of blood vessels. In contrast, oxygen and nutrients diffuse across the blood vessel layers and enter interstitial fluid, which carries oxygen and nutrients to the target cells, and carbon dioxide and wastes in the opposite direction. The other component of the circulatory system, the lymphatic system, is open.

Arteries

Oxygenated blood enters the systemic circulation when leaving the left ventricle, through the aortic semilunar valve. The first part of the systemic circulation is the aorta, a massive and thick-walled artery. The aorta arches and branches into major arteries to the upper body before passing through the diaphragm, where it branches further into arteries which supply the lower parts of the body.

Capillaries

Arteries branch into small passages called arterioles and then into the capillaries. The capillaries merge to bring blood into the venous system.

Veins

After their passage through body tissues, capillaries merge once again into venules, which continue to merge into veins. The venous system finally coalesces into two major veins: the superior vena cava (roughly speaking draining the areas above the heart) and the inferior vena cava (roughly speaking from areas below the heart). These two great vessels empty into the right atrium of the heart.

Coronary Vessels

The heart itself is supplied with oxygen and nutrients through a small "loop" of the systemic circulation.

Portal Veins

The general rule is that arteries from the heart branch out into capillaries, which collect into veins leading back to the heart. Portal veins are a slight exception to this. In humans the only significant example is the hepatic portal vein which combines from capillaries around the gut where the blood absorbs the various products of digestion; rather than leading directly back to the heart, the hepatic portal vein branches into a second capillary system in the liver.

Heart

The heart pumps oxygenated blood to the body and deoxygenated blood to the lungs. In the human heart there is one atrium and one ventricle for each circulation, and with both a systemic and a pulmonary circulation there are four chambers in total: left atrium, left ventricle, right atrium and right ventricle.

The right atrium is the upper chamber of the right side of the heart. The blood that is returned to the right atrium is deoxygenated (poor in oxygen) and passed into the right ventricle to be pumped through the pulmonary artery to the lungs for re-oxygenation and removal of carbon dioxide. The left atrium receives newly oxygenated blood from the lungs as well as the pulmonary vein which is passed into the strong left ventricle to be pumped through the aorta to the different organs of the body.

The coronary circulation system provides a blood supply to the heart muscle itself. The coronary circulation begins near the origin of the aorta by two arteries: the right coronary artery and the left coronary artery. After nourishing the heart muscle, blood returns through the coronary veins into the coronary sinus and from this one into the right atrium.

Back flow of blood through its opening during atrial systole is prevented by the Thebesian valve. The smallest cardiac veins drain directly into the heart chambers.

Pulmonary Circulation

Pulmonary circulation is the portion of the cardiovascular system which carries deoxygenated blood away from the heart, to the lungs, and returns oxygenated (oxygen-rich) blood back to the heart. The term pulmonary circulation is readily paired and contrasted with the systemic circulation. A separate system known as the bronchial circulation supplies blood to the tissue of the larger airways of the lung.

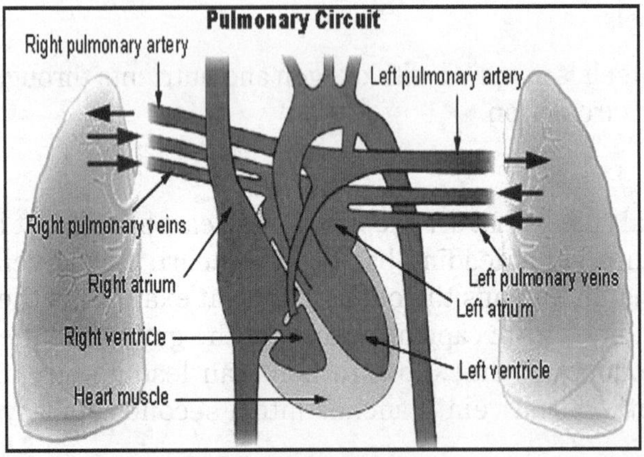

Fig. Diagram of Pulmonary Circulation.

Route

Pulmonary circulation is the movement of blood from the heart, to the lungs, and back to the heart again. Deoxygenated blood leaves the heart, goes to the lungs, and then re-enters the heart; Deoxygenated blood leaves through the right ventricle through the pulmonary artery. From the right atrium, the blood is pumped through the tricuspid valve (or right atrioventricular valve), into the right ventricle. Blood is then pumped from the right ventricle through the pulmonary valve and into the pulmonary trunk of the pulmonary artery.

- Arteries: From the right ventricle, blood is pumped through the pulmonary semi-lunar valve into the left and right pulmonary arteries (one for each lung) and travels through the lungs.
- Lungs: The pulmonary arteries carry deoxygenated blood to the lungs, where it releases carbon dioxide and picks up oxygen during respiration. Arteries are further divided in to very fine branches called the capillaries. In structure the capillaries are very thin walled. Their function is to assist in the carrying of blood to all cells of the body. The pulmonary vein returns oxygenated blood to the left atrium of the heart.
- Veins: The oxygenated blood then leaves the lungs through pulmonary veins, which return it to the left heart, completing the pulmonary cycle. This blood then enters the left atrium, which pumps it through the bicuspid valve, also called the mitral or left atrioventricular valve, into the left ventricle. From the left ventricle the blood passes through the aortic valve to the aorta. The blood is then distributed to the body through the systemic circulation before returning again to the pulmonary circulation.

History

Pulmonary circulation was first described by Ibn al-Nafis in his Commentary on Anatomy in Avicenna's Canon (1242). Ibn al-Nafis described pulmonary circulation as:

- "The blood from the right chamber of the heart must arrive at the left chamber but there is no direct pathway between them. The thick septum of the heart is not perforated and does not have visible pores as some people thought or invisible pores as Galen thought. The blood from the right chamber must flow through the vena arteriosa (pulmonary artery) to the lungs, spread through its substances, be mingled there with air, pass through the arteria venosa (pulmonary vein) to reach the left chamber of the heart and there form the vital spirit..."

It was later described by Michael Servetus in the "Manuscript of Paris" (near 1546, never published) and later published in his Christianismi Restitutio (1553). Since it was a theology work condemned by most of the Christian factions of his time, the discovery remained mostly unknown until the dissections of William Harvey in 1616.

Embryonic

The pulmonary circulation loop is virtually bypassed in fetal circulation. The fetal lungs are collapsed, and blood passes from the right atrium directly into the left atrium through the foramen ovale: an open conduit between the paired atria, or the ductus arteriosus: a shunt between the pulmonary artery and the aorta. When the lungs expand at birth, the pulmonary pressure drops and blood is drawn from the right atrium into the right ventricle and through the pulmonary circuit. Over the course of several months, the foramen ovale closes, leaving a shallow depression known as the fossa ovalis in the adult heart.

The circulatory system of the lungs is the portion of the cardiovascular system in which oxygen-depleted blood is pumped away from the heart, via the pulmonary artery, to the lungs and returned, oxygenated, to the heart via the pulmonary vein.

Oxygen deprived blood from the superior and inferior vena cava enters the right atrium of the heart and flows through the tricuspid valve (right atrioventricular valve) into the right ventricle, from which it is then pumped through the pulmonary semilunar valve into the pulmonary artery to the lungs. Gas exchange occurs in the lungs, whereby CO_2 is released from the blood, and oxygen is absorbed. The pulmonary vein returns the now oxygen-rich blood to the left atrium.

A separate system known as the bronchial circulation supplies blood to the tissue of the larger airways of the lung.

SYSTEMIC CIRCULATION

The systemic circulation is the circulation of the blood to all parts of the body except the lungs. Systemic circulation is the portion of the cardiovascular system which transports oxygenated blood away from the heart through the aorta from the left ventricle where the blood has been previously deposited from pulmonary circulation, to the rest of the body, and returns oxygen-depleted blood back to the heart.

BRAIN

The brain has a dual blood supply that comes from arteries at its front and back. These are called the "anterior" and "posterior" circulation respectively. The anterior circulation arises from the internal carotid arteries and supplies the front of the brain. The posterior circulation arises from the vertebral arteries, and supplies the back of the brain and brainstem. The circulation from the front and the back join together (anastomise) at the Circle of Willis.

Cerebral Circulation

Cerebral circulation is the movement of blood through the network of blood vessels supplying the brain. The arteries deliver oxygenated blood, glucose and other nutrients to the brain and the veins carry deoxygenated blood back to the heart, removing carbon dioxide, lactic acid, and other metabolic products. Since the brain is very vulnerable to compromises in its blood supply, the cerebral circulatory system has many safeguards. Failure of these safeguards results in cerebrovascular accidents, commonly known as strokes. The amount of blood that the cerebral circulation carries is known as cerebral blood flow. The presence of gravitational fields or accelerations also determine variations in the movement and distribution of blood in the brain, such as when suspended upside-down. The following description is based on idealized human cerebral circulation. The pattern of circulation and its nomenclature vary between organisms.

Arterial Cerebral Circulation

The arterial cerebral circulation is normally divided into anterior cerebral circulation and posterior cerebral circulation. There are two main pairs of arteries that supply the cerebral arteries and the cerebrum: Internal carotid arteries and vertebral arteries.

The anterior and posterior cerebral circulations are interconnected via bilateral posterior communicating arteries. They are part of the Circle of Willis, which provides backup circulation to the brain. In case one of the supply arteries is occluded, the Circle of Willis provides interconnections between the anterior and the posterior cerebral circulation along the floor of the cerebral vault, providing blood to tissues that would otherwise become ischemic.

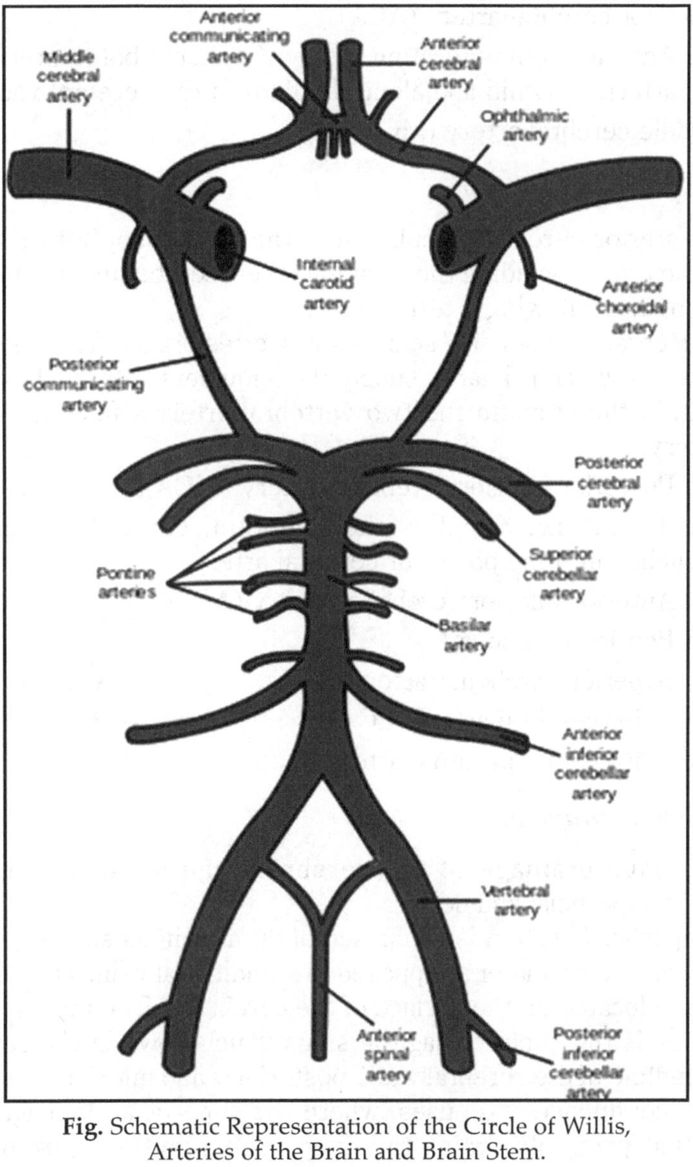

Fig. Schematic Representation of the Circle of Willis, Arteries of the Brain and Brain Stem.

Anterior Cerebral Circulation

The anterior cerebral circulation is the blood supply to the anterior portion of the brain. It is supplied by the following arteries:

- Internal carotid arteries: These large arteries are the left and right branches of the common carotid arteries in the neck which enter the skull, as opposed to the external carotid branches which supply the facial tissues. The internal carotid artery branches into the anterior cerebral artery and continues to form the middle cerebral artery

- Anterior cerebral artery (ACA)
  - Anterior communicating artery: Connects both anterior cerebral arteries, within and along the floor of the cerebral vault.
- Middle cerebral artery (MCA)

Posterior Cerebral Circulation

The posterior cerebral circulation is the blood supply to the posterior portion of the brain, including the occipital lobes, cerebellum and brainstem. It is supplied by the following arteries:

- Vertebral arteries: These smaller arteries branch from the subclavian arteries which primarily supply the shoulders, lateral chest and arms. Within the cranium the two vertebral arteries fuse into the basilar artery.
  - Posterior inferior cerebellar artery (PICA)
- Basilar artery: Supplies the midbrain, cerebellum, and usually branches into the posterior cerebral artery
  - Anterior inferior cerebellar artery (AICA)
  - Pontine branches
  - Superior cerebellar artery (SCA)
- Posterior cerebral artery (PCA)
- Posterior communicating artery

Cerebral Venous Drainage

The venous drainage of the cerebrum can be separated into two subdivisions: superficial and deep.

The superficial system is composed of dural venous sinuses, which have wall composed of dura mater as opposed to a traditional vein. The dural sinuses are, therefore located on the surface of the cerebrum. The most prominent of these sinuses is the superior sagittal sinus which flows in the sagittal plane under the midline of the cerebral vault, posteriorly and inferiorly to the torcula, forming the confluence of sinuses, where the superficial drainage joins with the sinus that primarily drains the deep venous system. From here, two transverse sinuses bifurcate and travel laterally and inferiorly in an S-shaped curve that form the sigmoid sinuses which go on to form the two jugular veins. In the neck, the jugular veins parallel the upward course of the carotid arteries and drain blood into the superior vena cava.

The deep venous drainage is primarily composed of traditional veins inside the deep structures of the brain, which join behind the midbrain to form the vein of Galen. This vein merges with the inferior sagittal sinus to form the straight sinus which then joins the superficial venous system mentioned above at the confluence of sinuses.

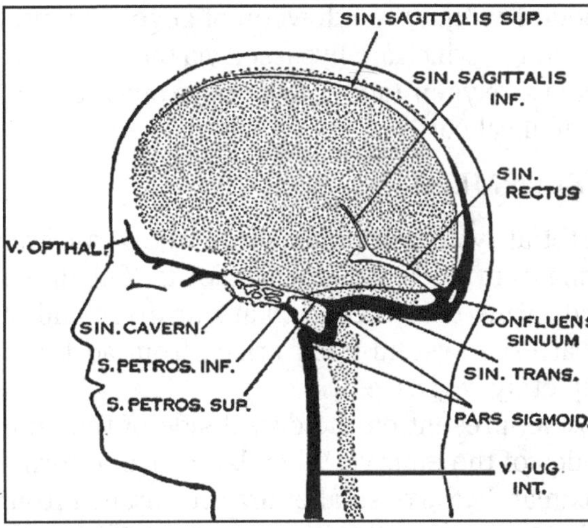

Fig. Dural Veins.

KIDNEYS

The renal circulation receives around 20% of the cardiac output. It branches from the abdominal aorta and returns blood to the ascending vena cava. It is the blood supply to the kidneys, and contains many specialized blood vessels.

LYMPHATIC SYSTEM

The lymphatic system is part of the circulatory system. It is a network of lymphatic vessels and lymph capillaries, lymph nodes and organs, and lymphatic tissues and circulating lymph. One of its major functions is to carry the lymph, draining and returning interstitial fluid back towards the heart for return to the cardiovascular system, by emptying into the lymphatic ducts. Its other main function is in the immune system.

PHYSIOLOGY

About 98.5% of the oxygen in a sample of arterial blood in a healthy human, breathing air at sea-level pressure, is chemically combined with hemoglobin molecules. About 1.5% is physically dissolved in the other blood liquids and not connected to hemoglobin. The hemoglobin molecule is the primary transporter of oxygen in mammals and many other species.

DEVELOPMENT

The development of the circulatory system starts with vasculogenesis in the embryo. The human arterial and venous systems develop from different areas in the embryo. The arterial system develops mainly from the aortic arches, six pairs of arches which develop on the upper part of the embryo. The venous system arises from three bilateral veins during weeks 4 – 8 of embryogenesis.

Fetal circulation begins within the 8th week of development. Fetal circulation does not include the lungs, which are bypassed via the truncus arteriosus. Before birth the fetus obtains oxygen (and nutrients) from the mother through the placenta and the umbilical cord.

ARTERIAL DEVELOPMENT

The human arterial system originates from the aortic arches and from the dorsal aortae starting from week 4 of embryonic life. The first and second aortic arches regress and forms only the maxillary arteries and stapedial arteries respectively. The arterial system itself arises from aortic arches 3, 4 and 6 (aortic arch 5 completely regresses).

The dorsal aortae, present on the dorsal side of the embryo, are initially present on both sides of the embryo. They later fuse to form the basis for the aorta itself. Approximately thirty smaller arteries branch from this at the back and sides. These branches form the intercostal arteries, arteries of the arms and legs, lumbar arteries and the lateral sacral arteries. Branches to the sides of the aorta will form the definitive renal, suprarenal and gonadal arteries. Finally, branches at the front of the aorta consist of the vitelline arteries and umbilical arteries. The vitelline arteries form the celiac, superior and inferior mesenteric arteries of the gastrointestinal tract. After birth, the umbilical arteries will form the internal iliac arteries.

VENOUS DEVELOPMENT

The human venous system develops mainly from the vitelline veins, the umbilical veins and the cardinal veins, all of which empty into the sinus venosus.

CLINICAL SIGNIFICANCE

Many diseases affect the circulatory system. This includes cardiovascular disease, affecting the cardiovascular system, and lymphatic disease affecting the lymphatic system. Cardiologists are medical professionals which specialise in the heart, and cardiothoracic surgeons specialise in operating on the heart and its surrounding areas. Vascular surgeons focus on other parts of the circulatory system.

CARDIOVASCULAR DISEASE

Diseases affecting the cardiovascular system are called cardiovascular disease.

Many of these diseases are called “lifestyle diseases” because they develop over time and are related to a person’s exercise habits, diet, whether they smoke, and other lifestyle choices a person makes. Atherosclerosis is the precursor to many of these diseases. It is where small atheromatous plaques build up in the walls of medium and large arteries. This may eventually grow or rupture to occlude the arteries. It is also a risk factor for acute coronary syndromes, which are diseases which are characterised by a sudden deficit of oxygenated blood to the heart tissue. Atherosclerosis is also associated with problems such as aneurysm formation or splitting ("dissection") of arteries.

Another major cardiovascular disease involves the creation of a clot, called a "thrombus". These can originate in veins or arteries. Deep venous thrombosis, which mostly occurs in the legs, is one cause of clots in the veins of the legs, particularly when a person has been stationary for a long time. These clots may embolise, meaning travel to another location in the body. The results of this may include pulmonary embolus, transient ischaemic attacks, or stroke.

Cardiovascular diseases may also be congenital in nature, such as heart defects or persistent fetal circulation, where the circulatory changes that are supposed to happen after birth do not. Not all congenital changes to the circulatory system are associated with diseases, a large number are anatomical variations.

MEASUREMENT TECHNIQUES

The function and health of the circulatory system and its parts are measured in a variety of manual and automated ways. These include simple methods such as those that are part of the cardiovascular examination, including the taking of a person's pulse as an indicator of a person's heart rate, the taking of blood pressure through a sphygmomanometer or the use of a stethoscope to listen to the heart for murmurs which may indicate problems with the heart's valves. An electrocardiogram can also be used to evaluate the way in which electricity is conducted through the heart.

Other more invasive means can also be used. A cannula or catheter inserted into an artery may be used to measure pulse pressure or pulmonary wedge pressures. Angiography, which involves injecting a dye into an artery to visualise an arterial tree, can be used in the heart (coronary angiography) or brain. At the same time as the arteries are visualised, blockages or narrowings may be fixed through the insertion of stents, and active bleeds may be managed by the insertion of coils. An MRI may be used to image arteries, called an MRI angiogram. For evaluation of the blood supply to the lungs a CT pulmonary angiogram may be used.

Ultrasound can also be used, particularly to identify the health of blood vessels, and a Doppler ultrasound of the carotid arteries or Doppler ultrasound of the lower limbs can be used to evaluate for narrowing of the carotid arteries or thrombus formation in the legs, respectively.

SURGERY

There are a number of surgical procedures performed on the circulatory system:

- Coronary artery bypass surgery

- Coronary stent used in angioplasty
- Vascular surgery
- Vein stripping
- Cosmetic procedures

Cardiovascular procedures are more likely to performed in the inpatient setting than in an ambulatory care setting; in the United States, only 28% of cardiovascular surgeries were performed in the ambulatory care setting.

OTHER VERTEBRATES

The circulatory systems of all vertebrates, as well as of annelids (for example, earthworms) and cephalopods (squids, octopuses and relatives) are closed, just as in humans. Still, the systems of fish, amphibians, reptiles, and birds show various stages of the evolution of the circulatory system.

In fish, the system has only one circuit, with the blood being pumped through the capillaries of the gills and on to the capillaries of the body tissues. This is known as single cycle circulation. The heart of fish is, therefore, only a single pump (consisting of two chambers).

In amphibians and most reptiles, a double circulatory system is used, but the heart is not always completely separated into two pumps. Amphibians have a three-chambered heart.

In reptiles, the ventricular septum of the heart is incomplete and the pulmonary artery is equipped with a sphincter muscle. This allows a second possible route of blood flow. Instead of blood flowing through the pulmonary artery to the lungs, the sphincter may be contracted to divert this blood flow through the incomplete ventricular septum into the left ventricle and out through the aorta. This means the blood flows from the capillaries to the heart and back to the capillaries instead of to the lungs. This process is useful to ectothermic (cold-blooded) animals in the regulation of their body temperature.

Birds and mammals show complete separation of the heart into two pumps, for a total of four heart chambers; it is thought that the four-chambered heart of birds evolved independently from that of mammals.

OPEN CIRCULATORY SYSTEM

The open circulatory system is a system in which a fluid in a cavity called the hemocoel bathes the organs directly with oxygen and nutrients and there is no distinction between blood and interstitial fluid; this combined fluid is called hemolymph or haemolymph. Muscular movements by the animal during locomotion can facilitate hemolymph movement, but diverting flow from one area to another is limited. When the heart relaxes, blood is drawn back toward the heart through open-ended pores (ostia).

Hemolymph fills all of the interior hemocoel of the body and surrounds all cells. Hemolymph is composed of water, inorganic salts (mostly sodium, chlorine, potassium, magnesium, and calcium), and organic compounds (mostly carbohydrates, proteins, and lipids). The primary oxygen transporter molecule is hemocyanin. There are free-floating cells, the hemocytes, within the hemolymph. They play a role in the arthropod immune system.

ABSENCE OF CIRCULATORY SYSTEM

Circulatory systems are absent in some animals, including flatworms (phylum Platyhelminthes). Their body cavity has no lining or enclosed fluid. Instead a muscular pharynx leads to an extensively branched digestive system that facilitates direct diffusion of nutrients to all cells. The flatworm's dorso-ventrally flattened body shape also restricts the distance of any cell from the digestive system or the exterior of the organism. Oxygen can diffuse from the surrounding water into the cells, and carbon dioxide can diffuse out. Consequently every cell is able to obtain nutrients, water and oxygen without the need of a transport system. Some animals, such as jellyfish, have more extensive branching from their gastrovascular cavity (which functions as both a place of digestion and a form of circulation), this branching allows for bodily fluids to reach the outer layers, since the digestion begins in the inner layers.

ASPECTS OF CIRCULATORY SYSTEM

ACDC (MEDICINE)

Arterial calcification due to deficiency of CD73 (ACDC) is a rare genetic disorder that causes calcium buildup in the arteries and joints of the hands and feet, and other areas below the waist. Although patients exhibiting these symptoms have been identified as early as 1914, this disorder had not been studied extensively until recently. The identification of the specific ACDC gene and mutations occurred in 2011. ACDC is caused by a mutation in the NT5E gene, which prevents calcium-removing agents from functioning. Patients afflicted with this mutation suffer from chronic pain, difficulty moving, and increased risk of cardiovascular problems. In experiments at the molecular level, treatment with adenosine or a phosphatase inhibitor reversed and prevented calcification, suggesting they could be used as possible treatment methods. There is currently no cure for ACDC, and patients have limited treatment options which focus primarily on removal of blood calcium and improving mobility.

Causation

In ACDC, the gene NT5E, which produces the enzyme CD73, is mutated. A mutation in NT5E causes CD73 to form incorrectly. A nonsense mutation, a single nucleotide insertion, and a missense mutation have all been found to produce essentially the same inactivation of CD73. Normally, CD73 binds to adenosine monophosphate, a nucleotide in DNA, and converts it to adenosine. In affected patients, little to no CD73 was functional. The calcification of cells is caused in part by a lack of pyrophosphate, which is broken down throughout the body by tissue-nonspecific alkaline phosphatase (TNAP). Without pyrophosphate, calcium phosphate crystals cannot be broken down. Since the inactive CD73 is unable to produce adenosine, which inhibits TNAP, there is an increase in TNAP levels and a decrease in pyrophosphate levels.

Treatment

The gene NT5E is related to the gene ENPP1, which when mutated is known to cause arterial calcification in infants. Treatments for the ENPP1 mutation, such as bisphosphonates, which are similar to phosphatase, and adenosine reuptake inhibitors, provide possible routes of treatment for the NT5E mutation, as the mechanism of both disorders are very similar.

AORTIC ORIFICE

The aortic orifice, (aortic opening) is a circular opening, in front and to the right of the left atrioventricular orifice, from which it is separated by the anterior cusp of the bicuspid valve. It is guarded by the aortic semilunar valve. The portion of the ventricle immediately below the aortic orifice is termed the aortic vestibule, and has fibrous instead of muscular walls.

BASILIC VEIN

The basilic vein is a large superficial vein of the upper limb that helps drain parts of the hand and forearm. It originates on the medial (ulnar) side of the dorsal venous network of the hand and travels up the base of the forearm, where its course is generally visible through the skin as it travels in the subcutaneous fat and fascia lying superficial to the muscles.

Near the region anterior to the cubital fossa, in the bend of the elbow joint, the basilic vein usually connects with the other large superficial vein of the upper extremity, the cephalic vein, via the median cubital vein (or median basilic vein). The layout of superficial veins in the forearm is highly variable from person to person, and there is a profuse network of unnamed superficial veins that the basilic vein communicates with.

As it ascends the medial side of the biceps in the arm proper (between the elbow and shoulder), the basilic vein normally perforates the brachial fascia (deep fascia) above the medial epicondyle, or even as high as mid-arm. There, around the lower border of the teres major muscle, the anterior and posterior circumflex humeral veins feed into it, just before it joins the brachial veins to form the axillary vein. Along with other superficial veins in the forearm, the basilic vein is an acceptable site for venipuncture. Nevertheless, IV nurses sometimes refer to the basilic vein as the "virgin vein," since with the arm typically supinated during phlebotomy the basilic vein below the elbow becomes awkward to access, and is therefore infrequently used. Vascular surgeons sometimes utilize the basilic vein to create an AV (arteriovenous) fistula or AV graft for hemodialysis access in patients with renal failure.

CARDIOVASCULAR PHYSIOLOGY

Cardiovascular physiology is the study of the circulatory system, specifically addressing the physiology of the heart ("cardio") and blood vessels ("vascular").These subjects are sometimes addressed separately, under the names cardiac physiology and circulatory physiology. Although the different aspects of cardiovascular physiology are closely interrelated, the subject is still usually divided into several subtopics.

Heart

- Cardiac output (= heart rate * stroke volume. Can also be calculated with Fick principle.)
  - Stroke volume (= end-diastolic volume - end-systolic volume)
  - Ejection fraction (= stroke volume/ end-diastolic volume)
  - ((Cardiac Output)) is mathematically ‘ to ((Systole))
  - Inotropic, chronotropic, and dromotropic states
  - Cardiac input (= heart rate * suction volume Can be calculated by inverting terms in Fick principle)
  - Suction volume (= end-systolic volume + end-diastolic volume)
  - Injection fraction (=suction volume/ end-systolic volume)
  - Cardiac input is mathematically ‘ to (Diastole))
- Electrical conduction system of the heart
  - Electrocardiogram
  - Cardiac marker
  - Cardiacs action potential
- Frank–Starling law of the heart
- Wiggers diagram
- Pressure volume diagram

Regulation of Blood Pressure

- Baroreceptor
- Baroreflex
- Renin-angiotensin system

  - Renin
  - Angiotensin
- Juxtaglomerular apparatus
- Aortic body and carotid body
- Autoregulation
  - Cerebral Autoregulation

Hemodynamics

Under most circumstances, the body attempts to maintain a steady mean arterial pressure.

When there is a major and immediate decrease (such as that due to hemorrhage or standing up), the body can increase the following:

- Heart
- Total peripheral resistance (primarily due to vasoconstriction of arteries)
- Inotropic state

In turn, this can have a significant impact upon several other variables:

- Stroke volume
- Cardiac output
- Pressure
  - Pulse pressure (systolic pressure - diastolic pressure)
  - Mean arterial pressure (usually approximated with diastolic pressure + 1/3 pulse pressure)
  - Central venous pressure

Regional Circulation

Name of circulation% of cardiac output Auto regulation Perfusion Commentspulmonary circulation100% (deoxygenated)Vasoconstriction in response to hypoxiacerebral circulation15%highunder-perfusedFixed volume means intolerance of high pressure. Minimal ability to use anaerobic respirationcoronary circulation5%highunder-perfusedMinimal ability to use anaerobic respiration.

Blood flow through the left coronary artery is at a maximum during diastole (in contrast to the rest of systemic circulation, which has a maximum blood flow during systole.)splanchnic circulation15%lowFlow increases during digestion.hepatic circulation15%Part of portal venous system, so oncotic pressure is very lowrenal circulation25%highover-perfusedMaintains glomerular filtration rateskeletal muscular circulation17%Perfusion increases dramatically during exercise.cutaneous circulation2%over-perfusedCrucial in thermoregulation. Significant ability to use anaerobic respiration

CEPHALIC VEIN

In human anatomy, the cephalic vein (also known as the antecubital vein)is a superficial vein of the upper limb. It communicates with the basilic vein via the median cubital vein at the elbow and is located in the superficial fascia along the anterolateral surface of the biceps brachii muscle. Superiorly the cephalic vein passes between the deltoid and pectoralis major muscles (deltopectoral groove) and through the deltopectoral triangle, where it empties into the axillary vein.

The cephalic vein is often visible through the skin, and its location in the deltopectoral groove is fairly consistent, making this site a good candidate for venous access. Permanent pacemaker leads are often placed in the cephalic vein in the deltopectoral groove. The vein may be used for intravenous access, and is sometimes referred to as the 'House-man's Friend' because a large bore cannula may be easily placed. However, the cannulation of a vein as close to the radial nerve as the cephalic vein can sometimes lead to nerve damage. Ordinarily the term cephalic refers to anatomy of the head. When Persian physician Ibn Sînâ's Canon was translated into medieval Latin, cephalic was mistakenly chosen to render the Arabic term al-kífal, meaning "outer".

ENDOTHELIAL ACTIVATION

Endothelial activation is a proinflammatory and procoagulant state of the endothelial cells lining the lumen of blood vessels. It is most characterized by an increase in interactions with white blood cells (leukocytes), and it is associated with the early states of atherosclerosis and sepsis, among others. It is also implicated in the formation of deep vein thrombosis. As a result of activation, enthothelium releases Weibel–Palade bodies.

HUMORAL FACTOR

Humoral factors are factors that are transported by the circulatory system, that is, in blood, and include:

- Humoral immunity factors in the immune system
- Hormones in the endocrine system

Humoral Immunity

Humoral immunity, also called the antibody-mediated beta cellularis immune system, is the aspect of immunity that is mediated by macromolecules (as opposed to cell-mediated immunity) found in extracellular fluids such as secreted antibodies, complement proteins and certain antimicrobial peptides. Humoral immunity is so named because it involves substances found in the humours, or body fluids. The study of the molecular and cellular components that form the immune system, including their function and interaction, is the central science of immunology. The immune system is divided into a more primitive innate immune system, and acquired or adaptive immune system of vertebrates, each of which contains humoral and cellular components.

Humoral immunity refers to antibody production and the accessory processes that accompany it, including: Th2 activation and cytokine production, germinal center formation and isotype switching, affinity maturation and memory cell generation. It also refers to the effector functions of antibodies, which include pathogen and toxin neutralization, classical complement activation, and opsonin promotion of phagocytosis and pathogen elimination.

Major discoveries in the study of humoral immunity		
Substance	**Activity**	**Discovery**
Alexin(s) Complement	Soluble components in the serum that are capable of killing microorganisms	Buchner (1890), Ehrlich (1892)
Antitoxins	Substances in the serum that can neutralize the activity of toxins, enabling passive immunization	von Behring and Kitasato (1890)
Bacteriolysins	Serum substances that work with the complement proteins to induce bacterial lysis	Richard Pfeiffer (1895)
Bacterial agglutinins and precipitins	Serum substances that agglutinate bacteria and precipitate bacterial toxins	von Gruber and Durham (1896), Kraus (1897)
Hemolysins	Serum substances that work with complement to lyse red blood cells	Belfanti and Carbone (1898) Jules Bordet (1899)
Opsonins	Serum substances that coat the outer membrane of foreign substances and enhance the rate of phagocytosis by macrophages	Wright and Douglas (1903)
Antibody	Formation (1900), antigen-antibody binding hypothesis (1938), produced by B cells (1948), structure (1972), immunoglobulin genes (1976)	Founder: P Ehrlich

The concept of humoral immunity developed based on analysis of antibacterial activity of the serum components. Hans Buchner is credited with the development of the humoral theory. In 1890 he described alexins, or "protective substances", which exist in the blood serum and other bodily fluid and are capable of killing microorganisms. Alexins, later redefined "complement" by Paul Ehrlich, were shown to be the soluble components of the innate response that lead to a combination of cellular and humoral immunity, and bridged the features of innate and acquired immunity.

Following the 1888 discovery of the bacteria that cause diphtheria and tetanus, Emil von Behring and Kitasato Shibasaburô showed that disease need not be caused by microorganisms themselves. They discovered that cell-free filtrates were sufficient to cause disease. In 1890, filtrates of diphtheria, later named diphtheria toxins, were used to vaccinate animals in an attempt to demonstrate that immunized serum contained an antitoxin that could neutralize the activity of the toxin and could transfer immunity to non-immune animals. In 1897, Paul Ehrlich showed that antibodies form against the plant toxins ricin and abrin, and proposed that these antibodies are responsible for immunity. Ehrlich, with his friend Emil von Behring, went on to develop the diphtheria antitoxin, which became the first major success of modern immunotherapy. The presence and specificity of compatibility antibodies became the major tool for standardizing the state of immunity and identifying the presence of previous infections.

Complement System

The complement system is a biochemical cascade of the innate immune system that helps clear pathogens from an organism. It is derived from many small blood plasma proteins that work together to disrupt the target cell's plasma membrane leading to cytolysis of the cell. The complement system consists of more than 35 soluble and cell-bound proteins, 12 of which are directly involved in the complement pathways. The complement system is involved in the activities of both innate immunity and acquired immunity.

Activation of this system leads to cytolysis, chemotaxis, opsonization, immune clearance, and inflammation, as well as the marking of pathogens for phagocytosis. The proteins account for 5% of the serum globulin fraction. Most of these proteins circulate as zymogens, which are inactive until proteolytic cleavage.

Three biochemical pathways activate the complement system: the classical complement pathway, the alternate complement pathway, and the mannose-binding lectin pathway. The classical complement pathway typically requires antibodies for activation and is a specific immune response, while the alternate pathway can be activated without the presence of antibodies and is considered a non-specific immune response. Antibodies, in particular the IgG1 class, can also "fix" complement.

B cells

The principal function of B cells is to make antibodies against soluble antigens. B cell recognition of antigen is not the only element necessary for B cell activation (a combination of clonal proliferation and the terminal differentiation into plasma cells). Naïve B cells can be activated in a T-cell dependent or independent manner, but two signals are always required to initiate activation.

B cell activation depends on one of three mechanisms: Type 1 T cell-independent (polyclonal) activation, Type 2 T cell-independent activation (in which mature B cells respond to highly repetitive structures causing cross-linking of the B cell receptors on the surface of B cells), and T cell-dependent activation. During T cell-dependent activation, an antigen presenting cell (APC) presents a processed antigen to a T helper cell (T_h), priming it. When a B cell processes and presents the same antigen to the primed T_h cell, the T cell releases cytokines that activate the B cell.

Antibodies

Immunoglobulins are glycoproteins in the immunoglobulin superfamily that function as antibodies. The terms antibody and immunoglobulin are often used interchangeably. They are found in the blood and tissue fluids, as well as many secretions. In structure, they are large Y-shaped globular proteins. In mammals there are five types of antibody: IgA, IgD, IgE, IgG, and IgM. Each immunoglobulin class differs in its biological properties and has evolved to deal with different antigens. Antibodies are synthesized and secreted by plasma cells that are derived from the B cells of the immune system.

An antibody is used by the acquired immune system to identify and neutralize foreign objects like bacteria and viruses. Each antibody recognizes a specific antigen unique to its target. By binding their specific antigens, antibodies can cause agglutination and precipitation of antibody-antigen products, prime for phagocytosis by macrophages and other cells, block viral receptors, and stimulate other immune responses, such as the complement pathway.

An incompatible blood transfusion causes a transfusion reaction, which is mediated by the humoral immune response. This type of reaction, called an acute hemolytic reaction, results in the rapid destruction (hemolysis) of the donor red blood cells by host antibodies. The cause is usually a clerical error, such as the wrong unit of blood being given to the wrong patient. The symptoms are fever and chills, sometimes with back pain and pink or red urine (hemoglobinuria). The major complication is that hemoglobin released by the destruction of red blood cells can cause acute renal failure.

Hormone

A hormone (from Greek ρμη, "impetus") is any member of a class of signaling molecules produced by glands in multicellular organisms that are transported by the circulatory system to target distant organs to regulate physiology and behaviour. Hormones have diverse chemical structures including eicosanoids, steroids, amino acid derivatives, peptides, and proteins. The glands that secrete hormones comprise the endocrine signaling system. The term hormone is sometimes extended to include chemicals produced by cells that affect the same cell (autocrine or intracrine signalling) or nearby cells (paracrine signalling).

Hormones are used to communicate between organs and tissues to regulate physiological and behavioral activities, such as digestion, metabolism, respiration, tissue function, sensory perception, sleep, excretion, lactation, stress, growth and development, movement, reproduction, and mood. Hormones affect distant cells by binding to specific receptor proteins in the target cell resulting in a change in cell function. When a hormone binds to the receptor, it results in the activation of a signal transduction pathway. This may lead to cell type-specific responses that include rapid non-genomic effects or slower genomic responses where the hormones acting through their receptors activate gene transcription resulting in increased expression of target proteins.

Hormone synthesis may occur in specific tissues of endocrine glands or in other specialized cells. Hormone synthesis occurs in response to specific biochemical signals induced by a wide range of regulatory systems. For instance, ionized calcium concentration affects PTH synthesis, whereas glucose concentration affects insulin synthesis. Regulation of hormone synthesis of gonadal, adrenal, and thyroid hormones is often dependent on complex sets of direct influence and feedback interactions involving the hypothalamic-pituitary-adrenal (HPA), -gonadal (HPG), and -thyroid (HPT) axes.

Upon secretion, certain hormones, including protein hormones and catecholamines, are water-soluble and are thus readily transported through the circulatory system. Other hormones, including steroid and thyroid hormones, are lipid-soluble; to allow for their widespread distribution, these hormones must bond to carrier plasma glycoproteins (e.g., thyroxine-binding globulin (TBG)) to form ligand-protein complexes. Some hormones are completely active when released into the bloodstream (as is the case for insulin and growth hormones), while others must be activated in specific cells through a series of activation steps that are commonly highly regulated. The endocrine system secretes hormones directly into the bloodstream typically into fenestrated capillaries, whereas the exocrine system secretes its hormones indirectly using ducts. Hormones with paracrine function diffuse through the interstitial spaces to nearby target tissue.

Hormonal signaling involves the following steps:

1. Biosynthesis of a particular hormone in a particular tissue
2. Storage and secretion of the hormone
3. Transport of the hormone to the target cell(s)
4. Recognition of the hormone by an associated cell membrane or intracellular receptor protein

5. Relay and amplification of the received hormonal signal via a signal transduction process: This then leads to a cellular response. The reaction of the target cells may then be recognized by the original hormone-producing cells, leading to a down-regulation in hormone production. This is an example of a homeostatic negative feedback loop.
6. Breakdown of the hormone.

Hormone cells are typically of a specialized cell type, residing within a particular endocrine gland, such as the thyroid gland, ovaries, and testes. Hormones exit their cell of origin via exocytosis or another means of membrane transport. The hierarchical model is an oversimplification of the hormonal signaling process. Cellular recipients of a particular hormonal signal may be one of several cell types that reside within a number of different tissues, as is the case for insulin, which triggers a diverse range of systemic physiological effects. Different tissue types may also respond differently to the same hormonal signal.

Regulation

The rate of hormone biosynthesis and secretion is often regulated by a homeostatic negative feedback control mechanism. Such a mechanism depends on factors that influence the metabolism and excretion of hormones. Thus, higher hormone concentration alone cannot trigger the negative feedback mechanism. Negative feedback must be triggered by overproduction of an "effect" of the hormone.

Hormone secretion can be stimulated and inhibited by:

- Other hormones (stimulating- or releasing -hormones)
- Plasma concentrations of ions or nutrients, as well as binding globulins
- Neurons and mental activity
- Environmental changes, e.g., of light or temperature

One special group of hormones is the tropic hormones that stimulate the hormone production of other endocrine glands. For example, thyroid-stimulating hormone (TSH) causes growth and increased activity of another endocrine gland, the thyroid, which increases output of thyroid hormones. To release active hormones quickly into the circulation, hormone biosynthetic cells may produce and store biologically inactive hormones in the form of pre- or prohormones. These can then be quickly converted into their active hormone form in response to a particular stimulus. Eicosanoids are considered to act as local hormones.

Receptors

Most hormones initiate a cellular response by initially binding to either cell membrane associated or intracellular receptors. A cell may have several different receptor types that recognize the same hormone but activate different signal transduction pathways, or a cell may have several different receptors that recognize different hormones and activate the same biochemical pathway.

Receptors for most peptide as well as many eicosanoid hormones are embedded in the plasma membrane at the surface of the cell and the majority of these receptors belong to the G protein-coupled receptor (GPCR) class of seven alpha helix transmembrane proteins. The interaction of hormone and receptor typically triggers a cascade of secondary effects within the cytoplasm of the cell, often involving phosphorylation or dephosphorylation of various other cytoplasmic proteins, changes in ion channel permeability, or increased concentrations of intracellular molecules that may act as secondary messengers (e.g., cyclic AMP). Some protein hormones also interact with intracellular receptors located in the cytoplasm or nucleus by an intracrine mechanism.

For steroid or thyroid hormones, their receptors are located inside the cell within the cytoplasm of the target cell. These receptors belong to the nuclear receptor family of ligand-activated transcription factors. To bind their receptors, these hormones must first cross the cell membrane. They can do so because they are lipid-soluble. The combined hormone-receptor complex then moves across the nuclear membrane into the nucleus of the cell, where it binds to specific DNA sequences, regulating the expression of certain genes, and thereby increasing the levels of the proteins encoded by these genes. However, it has been shown that not all steroid receptors are located inside the cell. Some are associated with the plasma membrane.

Effects

A variety of exogenous chemical compounds, both natural and synthetic, have hormone-like effects on both humans and wildlife. Their interference with the synthesis, secretion, transport, binding, action, or elimination of natural hormones in the body can change the homeostasis, reproduction, development, and/or behavior, similar to endogenously produced hormones.

Hormones have the following effects on the body:

- Stimulation or inhibition of growth
- Wake-sleep cycle and other circadian rhythms
- Mood swings
- Induction or suppression of apoptosis (programmed cell death)
- Activation or inhibition of the immune system
- Regulation of metabolism
- Preparation of the body for mating, fighting, fleeing, and other activity
- Preparation of the body for a new phase of life, such as puberty, parenting, and menopause
- Control of the reproductive cycle

- Hunger cravings
- Sexual arousal

A hormone may also regulate the production and release of other hormones. Hormone signals control the internal environment of the body through homeostasis.

Chemical Classes

As hormones are defined functionally, not structurally, they may have diverse chemical structures. Hormones occur in multicellular organisms (plants, animals, fungi, brown algae and red algae). These compounds occur also in unicellular organisms, and may act as signaling molecules, but there is no consensus if, in this case, they can be called hormones.

Animal

Vertebrate hormones fall into four main chemical classes:

- Amino acid derived – Examples include melatonin and thyroxine.
- Peptides, polypeptides and proteins. – Small peptide hormones include TRH and vasopressin. Peptides composed of scores or hundreds of amino acids are referred to as proteins. Examples of protein hormones include insulin and growth hormone. More complex protein hormones bear carbohydrate side-chains and are called glycoprotein hormones. Luteinizing hormone, follicle-stimulating hormone and thyroid-stimulating hormone are examples of glycoprotein hormones.
- Eicosanoids – hormones derive from lipids such as arachidonic acid, lipoxins and prostaglandins.
- Steroid – Examples of steroid hormones include the sex hormones estradiol and testosterone as well as the stress hormone cortisol.

Compared with vertebrate, insects and crustaceans possess a number of structurally unusual hormones such as the juvenile hormone, a sesquiterpenoid.

Plant

Plant hormones include abscisic acid, auxin, cytokinin, ethylene, and gibberellin.

Therapeutic use

Many hormones and their analogues are used as medication. The most commonly prescribed hormones are estrogens and progestogens (as methods of hormonal contraception and as HRT), thyroxine (as levothyroxine, for hypothyroidism) and steroids (for autoimmune diseases and several respiratory disorders). Insulin is used by many diabetics. Local preparations for use in otolaryngology often contain pharmacologic equivalents of adrenaline, while steroid and vitamin D creams are used extensively in dermatological practice. A "pharmacologic dose" or "supraphysiological dose" of a hormone is a medical usage referring to an amount of a hormone far greater than naturally occurs in a healthy body. The effects of pharmacologic doses of hormones may be different from responses to naturally occurring amounts and may be therapeutically useful, though not without potentially adverse side effects. An example is the ability of pharmacologic doses of glucocorticoids to suppress inflammation.

Hormone-Behavior Iinteractions

At the neurological level, behavior can be inferred based on: hormone concentrations; hormone-release patterns; the numbers and locations of hormone receptors; and the efficiency of hormone receptors for those involved in gene transcription. Not only do hormones influence behavior, but also behavior and the environment influence hormones. Thus, a feedback loop is formed. For example, behavior can affect hormones, which in turn can affect behavior, which in turn can affect hormones, and so on.

Three broad stages of reasoning may be used when determining hormone-behavior interactions:

- The frequency of occurrence of a hormonally dependent behavior should correspond to that of its hormonal source
- A hormonally dependent behavior is not expected if the hormonal source (or its types of action) is non-existent
- The reintroduction of a missing behaviorally dependent hormonal source (or its types of action) is expected to bring back the absent behavior

Comparison with Neurotransmitters

There are various clear distinctions between hormones and neurotransmitters:

- A hormone can perform functions over a larger spatial and temporal scale than can a neurotransmitter.
- Hormonal signals can travel virtually anywhere in the circulatory system, whereas neural signals are restricted to pre-existing nerve tracts
- Assuming the travel distance is equivalent, neural signals can be transmitted much more quickly (in the range of milliseconds) than can hormonal signals (in the range of seconds, minutes, or hours). Neural signals can be sent at speeds up to 100 meters per second.
- Neural signalling is an all-or-nothing (digital) action, whereas hormonal signalling is an action that can be continuously variable as dependent upon hormone concentration

MICROCIRCULATION

Microcirculation is the circulation of the blood in the smallest blood vessels, present in the vasculature embedded within organ tissues. This contrasts with macrocirculation, which is the circulation of blood to and from the organs. The microcirculation is composed of terminal arterioles, capillaries, and venules that drain capillary blood. The vessels on the arterial side of the microcirculation are called the arterioles, which are well innervated, are surrounded by smooth muscle cells, and are 10-100 μm in diameter. Arterioles carry the blood to the capillaries, which are not innervated, have no smooth muscle, and are about 5-8 μm in diameter. Blood flows out of the capillaries into the venules, which have little smooth muscle and are 10-200 μm. The blood flows from the venules into the veins. In addition to these blood vessels, the microcirculation also includes lymphatic capillaries and collecting ducts. The main functions of the microcirculation are the delivery of oxygen and nutrients and the removal of carbon dioxide (CO_2). It also serves to regulate blood flow and tissue perfusion thereby affecting blood pressure and responses to inflammation which can include edema (swelling).

Most vessels of the microcirculation are lined by flattened cells of the endothelium and many of them are surrounded by contractile cells called pericytes. The endothelium provides a smooth surface for the flow of blood and regulates the movement of water and dissolved materials in the interstitial plasma between the blood and the tissues. The endothelium also produces molecules that discourage the blood from clotting unless there is a leak. Pericyte cells can contract and decrease the size of the arterioles and thereby regulate blood flow and blood pressure.

Sectors of Microcirculation

Microcirculation has three major sectors: pre-capillary (resistive), capillary (swap) and post-capillary (capacitive). In the pre-capillary sector, arterioles and precapillary sphincters participate. Their function is to regulate blood flow before it enters the capillaries and venules by the contraction and relaxation of the smooth muscle found on their walls. The second sector is the capillary sector, which is represented by the capillaries, where substance and gas exchange between blood and interstitial fluid takes place. Finally, the post-capillary sector is represented by the post-capillary venules, which are formed by a layer of endothelial cells that allow free movement of some substances.

Regulation of Microcirculation

The regulation of tissue perfusion occurs in microcirculation. There, arterioles control the flow of blood to the capillaries. Arterioles contract and relax, varying their diameter and vascular tone, as the vascular smooth muscle responds to diverse stimuli. Distension of the vessels due to increased blood pressure is a fundamental stimulus for muscle contraction in arteriolar walls. As a consequence, microcirculation blood flow remains constant despite of changes in systemic blood pressure. This mechanism is present in all tissues and organs of the human body.

In addition, the nervous system participates in the regulation of microcirculation. The sympathetic nervous system activates the smaller arterioles, including terminals. Noradrenaline and adrenaline have effects on alpha and beta adrenergic receptors. Other hormones (catecholamine, renin-angiotensin, vasopressin, and atrial natriuretic peptide) circulate in the bloodstream and can have an effect on the microcirculation causing vasodilation or vasoconstriction. Many hormones and neuropeptides are released together with classical neurotransmitters.

Arterioles respond to metabolic stimuli that are generated in the tissues. When tissue metabolism increases, catabolic products accumulate leading to vasodilation. The endothelium begins to control muscle tone and arteriolar blood flow tissue. Endothelial function in the circulation includes the activation and inactivation of circulating hormones and other plasma constituents. There are also synthesis and secretion of vasodilator and vasoconstrictor substances for modifying the width as necessary. Variations in the flow of blood that circulates by arterioles are capable of responses in endothelium.

Capillary Exchange

The term capillary exchange refers to all exchanges at microcirculatory level, most of which occurs in the capillaries. Sites where material exchange occurs between the blood and tissues are the capillaries, which branch out to increase the swap area, minimize the diffusion distance as well as maximize the surface area and the exchange time.

Approximately, seven percent of the body's blood is in the capillaries which continuously exchange substances with the liquid outside these blood vessels, called interstitial fluid. This dynamic displacement of materials between the interstitial fluid and the blood is named capillary exchange. These substances pass through capillaries through three different systems or mechanisms: diffusion, bulk flow, and transcytosis or vesicular transport. The liquid and solid exchanges that take place in the microvasculature particularly involve capillaries and post-capillary venules and collecting venules.

Capillary walls allow the free flow of almost every substance in plasma. The plasma proteins are the only exception, as they are too big to pass through. The minimum number of un-absorbable plasma proteins that exit capillaries enter lymphatic circulation for returning later on to those blood vessels. Those proteins which leave capillaries use the first capillary exchange mechanism and the process of diffusion, which is caused by kinetic motion of molecules.

Regulation of Capillary Exchange

These exchanges of substances are regulated by different mechanisms. These mechanisms work together and promote capillary exchange in the following way. First, molecules diffuse that are going to travel a short distance thanks to the capillary wall, the small diameter and the close proximity to each cell having a capillary. The short distance is important because the capillary diffusion rate decreases when the diffusion distance increases. Then, because of its large number (10-14 million capillaries), there is an incredible amount of surface area for exchange. However, this only has 5% of the total blood volume (250 ml 5000 ml). Finally, blood flows more slowly in the capillaries, given the extensive branching.

Diffusion

Diffusion is the first and most important mechanism that allows the flow of small molecules across capillaries. The process depends on the difference of gradients between the interstitium and blood, with molecules moving to low concentrated spaces from high concentrated ones. Glucose, amino acids, oxygen (O_2) and other molecules exit capillaries by diffusion to reach the organism's tissues. Contrarily, carbon dioxide (CO_2) and other wastes leave tissues and enter capillaries by the same process but in reverse. Diffusion through the capillary walls depends on the permeability of the wall to which exchange materials, this permeability depends on the endothelial cells forming the capillary walls which may be continuous, discontinuous, and fenestrated. The Starling equation is an equation that describes the roles of hydrostatic and osmotic forces (the so-called Starling forces) in the movement of fluid across capillary endothelium.

Bulk Flow

The second mechanism of capillary exchange is bulk flow. It is used by small, lipid-insoluble substances in order to cross. This movement depends on the physical characteristics of the capillaries. For example, continuous capillaries (tight structure) reduce bulk flow, fenestrated capillaries (perforated structure) increases bulk flow, and discontinuous capillaries (great intercellular gaps) enable bulk flow. In this case, the exchange of materials is determined by changes in pressure. When the flow of substances goes from the bloodstream or the capillary to the interstitial space or interstitium, the process is called filtration. This kind of movement is favored by blood hydrostatic pressure (BHP) and interstitial fluid osmotic pressure (IFOP). Otherwise, if the substances move from the interstitial fluid to the blood in capillaries, the process is called reabsorption. The pressures that favor this movement are blood colloid osmotic pressure (BCOP) and interstitial fluid hydrostatic pressure (IFHP). A substance will be filtrated or reabsorbed depending on the net filtration pressure (NFP).

This force is the result of the balance between hydrostatic (BHP and IFHP) and osmotic pressures (IFOP and BCOP). These pressures are known as the Starling forces. If the calculation of the NFP is positive then there will be filtration, but if it is negative then reabsorption will occur.

Transcytosis

The third capillary exchange mechanism is transcytosis, also called vesicular transport. By this process, blood substances move across the endothelial cells that compose the capillary structure. Finally, these materials exit by exocytosis, process in which vesicles go out from a cell to the interstitial space. Transcytosis is mainly used by large molecules that are lipid-insoluble, such as the insulin hormone. A minimum amount of substances cross by transcytosis. Once vesicles exit the capillaries, they go to the interstitium. Sometimes vesicles can merge with other vesicles, so their contents are mixed, or can directly go to a specific tissue. This material intermixed increases the functional capability of the vesicle.

PRECAPILLARY RESISTANCE

Precapillary resistance is the modulation of blood flow by capillaries through vasomotion, either opening (dilating) and letting blood pass through, or by constricting their lumens, reducing bloodflow through the capillary bed (occluding the passage of blood). It is not entirely clear how precapillary resistance is created in many parts of the body. Precapillary sphincters are smooth muscle structures that mediate the precapillary resistance in the mesenteric microcirculation.

PRECAPILLARY SPHINCTER

A precapillary sphincter is a band of smooth muscle that adjusts blood flow into capillaries mainly in the mesenteric microcirculation. At the point where each of the capillaries originates from an arteriole, a smooth muscle fiber encircles the capillary. This is called the precapillary sphincter. The sphincter can open and close the entrance to the capillary, by which contraction causes blood flow in a capillary to change as vasomotion occurs. The entire capillary bed may be bypassed by blood flow through arteriovenous anastomoses or through preferential flow through metarterioles. If the sphincter is damaged or cannot contract, blood can flow into the capillary bed at high pressures. When capillary pressures are high (as per gravity, etc.), fluid passes out of the capillaries into the interstitial space, and edema or fluid swelling is the result. Precapillary sphincters and metarterioles were discovered in the mesenteric circulation in the 1950s. Medical and physiological textbooks, such as Guyton, Boron and Fulton, etc. were quick to claim the existence of said sphincters and metarterioles all over the body, despite lack of evidence. At least since 1976 there has been considerable debate about the existence of precapillary sphincters and metarterioles. As of 2013 it is held that they are unique to the mesenteric circulation and some researchers have suggested the term precapillary resistance instead.

RETE MIRABILE

A rete mirabile (Latin for "wonderful net"; plural retia mirabilia) is a complex of arteries and veins lying very close to each other, found in some vertebrates, mainly warm-blooded ones. The rete mirabile utilizes countercurrent blood flow within the net (blood flowing in opposite directions) to act as a countercurrent exchanger. It exchanges heat, ions, or gases between vessel walls so that the two bloodstreams within the rete maintain a gradient with respect to temperature, or concentration of gases or solutes. This term was coined by Galen.

The effectiveness of retia is primarily determined by how readily the heat, ions, or gases can be exchanged. For a given length, they are most effective with respect to gases or heat, then small ions, and decreasingly so with respect to other substances.

The retia can provide for extremely efficient exchanges. In bluefin tuna, for example, nearly all of the metabolic heat in the venous blood is transferred to the arterial blood, thus conserving muscle temperature; that heat exchange approaches 99% efficiency.

In birds with webbed feet, retia mirabilia in the legs and feet transfer heat from the outgoing (hot) blood in the arteries to the incoming (cold) blood in the veins. The effect of this biological heat exchanger is that the internal temperature of the feet is much closer to the ambient temperature, thus reducing heat loss. Penguins have them also in the flippers and nasal passages.

Seabirds distill seawater using countercurrent exchange in a gland with a rete mirabile. The gland secretes highly concentrated brine stored near the nostrils above the beak. The bird then "sneezes" the brine out. As freshwater is not usually available in their environments, some seabirds, such as pelicans, petrels, albatrosses, gulls and terns, possess this gland, which allows them to drink the salty water from their environments while they are hundreds of miles away from land.

In some fish, a rete mirabile fills the swim bladder with oxygen, increasing the fish's buoyancy. A countercurrent exchange system is utilized between the venous and arterial capillaries. By lowering the pH levels in the venous capillaries, oxygen unbinds from blood hemoglobin. This causes an increase in venous blood oxygen partial pressure, allowing the oxygen to diffuse through the capillary membrane and into the arterial capillaries, where oxygen is still sequestered to hemoglobin. The cycle of diffusion continues until the partial pressure of oxygen in the arterial capillaries exceeds that in the swim bladder.

At this point, the dissolved oxygen in the arterial capillaries diffuses into the swim bladder via the gas gland.

The rete mirabile allows for an increase in muscle temperature in regions where this network of vein and arteries is found. The fish is able to thermoregulate certain areas of their body. Additionally, this increase in temperature leads to an increase in basal metabolic temperature. The fish is now able to split ATP at a higher rate and ultimately can swim faster.

The opah utilizes retia mirabilia to conserve heat, making it the only warm-blooded fish. Blood traveling through capillaries in the gills must carry cold blood due to their exposure to cold water, but retia mirabilia in the opah's gills are able to transfer heat from warm blood in arterioles coming from the heart that heats this colder blood in arterioles leaving the gills. The huge pectoral muscles of the opah, which generate most of the body heat, are thus able to control the temperature of the rest of the body.

In mammals, an elegant rete mirabile in the efferent arterioles of juxtamedullary glomeruli is important in maintaining the hypertonicity of the renal medulla. It is the hypertonicity of this zone, resorbing water osmotically from the renal collecting ducts as they exit the kidney, that makes possible the excretion of a hypertonic urine and maximum conservation of body water.

Vascular retia mirabilia are also found in the limbs of a range of mammals. These reduce the temperature in the extremities. Some of these probably function to prevent heat loss in cold conditions by reducing the temperature gradient between the limb and the environment. Others reduce the temperature of the testes increasing their productivity. In the neck of the dog, a rete mirabile protects the brain when the body overheats during hunting; the venous blood is cooled down by panting before entering the net.

Retia mirabilia also occur frequently in mammals that burrow, dive or have arboreal lifestyles that involve clinging with the limbs for lengthy periods. In these cases, the rete mirabile may lower limb temperature and therefore the metabolic requirement for oxygen and nutrients in the tissues of the limb. This would be desirable when the blood supply to the limb is limited as a result of a "diving reflex" diverting blood away from the limbs during diving or burrowing or because of postural restrictions on blood supply in the case of clinging, arboreal animals. In the last case, slow-moving arboreal mammals such as sloths, lorises and arboreal anteaters possess retia of the highly developed type known as vascular bundles. The structure and function of these mammalian retia mirabilia are reviewed by O'Dea (1990).

In giraffes, a rete mirabile in the neck equalizes blood pressure when the animal bends down to drink.

The ancient physician Galen mistakenly thought that humans also have a rete mirabile in the neck, apparently based on dissection of sheep and misidentifying the results with the human carotid sinus, and ascribed important properties to it; it fell to Berengario da Carpi first, and then to Vesalius to demonstrate the error.

TRANSPORT MAXIMUM

In physiology, transport maximum (alternatively Tm or T_{max}) refers to the point at which increases in concentration do not result in an increase in movement of a substance across a membrane. In renal physiology, the concept of transport maximum is often discussed in the context of glucose and PAH. For both substances (as with all substances), the quantity excreted can be determined with the following equation:

- excretion = (filtration + secretion) - reabsorption
- The proximal convoluted tubule of the nephron has protein channels that reabsorb glucose, and others that secrete para-aminohippuric acid (PAH). However, its ability to do so is proportionate to the channel proteins available for the transport.
- Glucose is not secreted, so excretion = filtration - reabsorption. Both filtration and reabsorption are directly proportional to the concentration of glucose in the plasma. However, while the maximum reabsorption is about 260 mg/min in healthy nephrons, filtration has effectively no limit (within reasonable physiological ranges.) Therefore, if the concentration rises above 260 mg/min, the body cannot retain all the glucose, leading to glucosuria.
- PAH is not reabsorbed and is secreted, so excretion = filtration + secretion. As with glucose, the transfer is at the proximal tubule, but in the opposite direction: from the peritubular capillaries to the lumen. At low levels, all the PAH is transferred, but at high levels, the transport maximum is reached, and the PAH takes longer to clear.

In practice, the transport maximum is not all-or-nothing. As the concentration approaches the transport maximum, some of the channels are overwhelmed before others are. For example, with glucose, some sugar appears in the urine at levels much lower than 300 mg/dL. The point at which the effects start to appear is called "threshold", and the difference between threshold and transport maximum is called "splay".

VASCULARITY

Vascularity, in bodybuilding, is the condition of having many highly visible, prominent and often extensively ramified superficial veins. The skin appears "thin" — sometimes virtually diaphanous — due to an extreme reduction of subcutaneous fat, allowing for maximum muscle definition.

Vascularity is enhanced by extremely low body fat (usually below 10%) and low retained water, as well as the muscle engorgement ("pump") and venous distension accentuated by the vigorous flexing and potentially hazardous

Valsalva effect which characterize competitive posing. Genetics and androgenic hormones will affect vascularity, as will ambient temperature. (And although some bodybuilders develop arterial hypertension from performance-enhancing substances and practices, "high" venous pressure — being an order of magnitude lower than that of arteries — neither causes nor is caused by vascularity.) Some bodybuilders use topical vasodilators to increase blood flow to the skin as well. Although historically controversial, nowadays vascularity is a highly sought after aesthetic for many male bodybuilders, but less so for female bodybuilders where the target aesthetic is relatively more towards aesthetic symmetry than extreme development.

Bodybuilders or athletes sometimes dehydrate themselves, a few days before a competition or show, to achieve this so-called "ripped" vascular look. Self-dehydration is not recommended by medical professionals, as the negative and sometimes tragic effects of the resultant water-electrolyte imbalances are well documented.

VASOMOTION

Vasomotion is the spontaneous oscillation in tone of blood vessels, independent of heart beat, innervation or respiration. While vasomotion was first observed by Jones in 1852, the complete mechanisms responsible for its generation and its physiological importance remain to be elucidated, however several hypotheses have been put forth.

Intracellular calcium (Ca^{2+}) concentration exhibits periodic oscillations in vascular smooth muscle cells. This is thought to result from Ca^{2+} release from intracellular stores, due to inositol triphosphate and ryanodine-sensitive channel activation. This activation has been shown to result in either Ca^{2+} "sparks", highly localized calcium increases, or "waves", global Ca^{2+} increase that propagates the length of the cell.

To allow vasomotion to occur, synchronization must occur between the individual oscillations, resulting in global calcium synchronization and vessel tone oscillation. Gap junctions are thought to play a large role in this synchronization, as application of gap junction blockers has been shown to abolish vasomotion, indicating a critical role. Due to regional variations in gap junction distribution and coupling (homocellular vs. heterocellular) several hypotheses have been suggested to account for vasomotion occurrence.

The "classic" mechanism of vasomotion generation is thought to be the voltage-dependent coupled model. In this model, high gap junction coupling is present between the vascular smooth muscle cells, the endothelial cells and the endothelial to vascular smooth muscle cells. An initial depolarizing current leads to the opening of the voltage-dependent calcium channels, ultimately resulting in synchronization of individual calcium levels. When patch clamp recordings are conducted, depolarization occurs in the endothlial layer at the same time as the underlying vascular smooth muscle. The cause of the initial depolarizing current, however, remains to be determined. Mathematical modeling has pointed to the existence of 2-4 independent non-linear oscillating systems interacting to produce vasomotion. It is possible that in order for vasomotion to be generated, these systems must pass a depolarizing threshold.

Several possible hypotheses have been advanced to explain vasomotion. Increased flow is one possibility; mathematical modeling has shown a vessel with an oscillating diameter to conduct more flow then a vessel with a static diameter. Vasomotion could also be a mechanism of increasing the reactivity of a blood vessel by avoiding the "latch state", a low ATP cycling state of prolonged force generation common in vascular smooth muscle. Finally, vasomotion has been shown to be altered in a variety of pathological situations, with vessels from both hypertensive and diabetic patients displaying altered flow patterns as compared to normotensive vessels.

WATERSHED AREA (MEDICAL)

Watershed area is the medical term referring to regions of the body that receive dual blood supply from the most distal branches of two large arteries, such as the splenic flexure of the large intestine.

During times of blockage of one of the arteries that supply the watershed area, such as in atherosclerosis, these regions are spared from ischemia by virtue of their dual supply. However, during times of systemic hypoperfusion, such as in disseminated intravascular coagulation or heart failure, these regions are particularly vulnerable to ischemia by virtue of the fact that they are supplied by the most distal branches of their arteries, and thus the least likely to receive sufficient blood.

Watershed areas are found in the brain, where areas are perfused by both the anterior and middle cerebral arteries, and in the intestines, where areas are perfused by both the superior and inferior mesenteric arteries (i.e., splenic flexure). Additionally, the sigmoid colon and rectum form a watershed zone with blood supply from inferior mesenteric, pudendal and iliac circulations. Hypoperfusion in watershed areas can lead to mural and mucosal infarction in the case of ischemic bowel disease. When watershed stroke occurs in the brain, it produces unique focal neurologic symptoms that aid clinicians in diagnosis and localization.

Bibliography

A Rajam: *Blood Sweat and Tears*, Niyogi Books, Delhi, 2013.

Ameeta Rathore: *Blood Ties*, Penguin Books, Delhi, 2001.

Bolton, S J: *Blood Harvest*, Random House India, Delhi, 2010.

Bradby, Tom: *Blood Money*, Random House India, Delhi, 2010.

Carter, Angela: *Bloody Chamber And Other Stories, The*, Random House India, Delhi, 2009.

Carter, Angela: *Bloody Chamber And Other Stories, The*, Random House India, Delhi, 1999.

Carter, Dean Vincent: *Blood Water*, Random House India, Delhi, 2009.

Conlon, Edward: *Blue Blood*, Random House India, Delhi, 2010.

Ellroy, James: *Blood's A Rover*, Random House India, Delhi, 2010.

Fleming, James: *Cold Blood*, Random House India, Delhi, 2009.

Fleming, James: *Cold Blood*, Random House India, Delhi, 2010.

Gerald Schumacher: *Bloody Business : Americas War Zone Contractors and the Occupation of Iraq*, Manas Publication, Delhi, 2007.

Holland, James: *Blood of Honour*, Random House India, Delhi, 2010.

James W. Jones: *Blood That Cries Out from the Earth : The Psychology of Religious Terrorism*, Oxford University Press, Delhi, 2008.

M J Akbar: *Blood Brothers : A Family Saga*, Roli Books, Delhi, 2006.

Mina, Denise: *Field Of Blood, The*, Random House India, Delhi, 2006.

Pearson, Mark: *Blood Work*, Random House India, Delhi, 2009.

Rice, Anne: *Blood Canticle*, Random House India, Delhi, 2010.

Russell, Craig: *Blood Eagle*, Random House India, Delhi, 2006.

S. Nihal Singh: *Blood and Sand : The West Asian Tragedy*, CBS Publishers, Delhi, 2003.

Sabatini, Rafael: *Captain Blood*, Random House India, Delhi, 2009.

Seymour, Gerald: *Field Of Blood*, Random House India, Delhi, 1999.

Tessa Gratton: *Blood Magic*, Random House, Delhi, 2011.

Index